I0463096

ALGEBRA EXAMPLES

BASIC FUNCTIONS 2

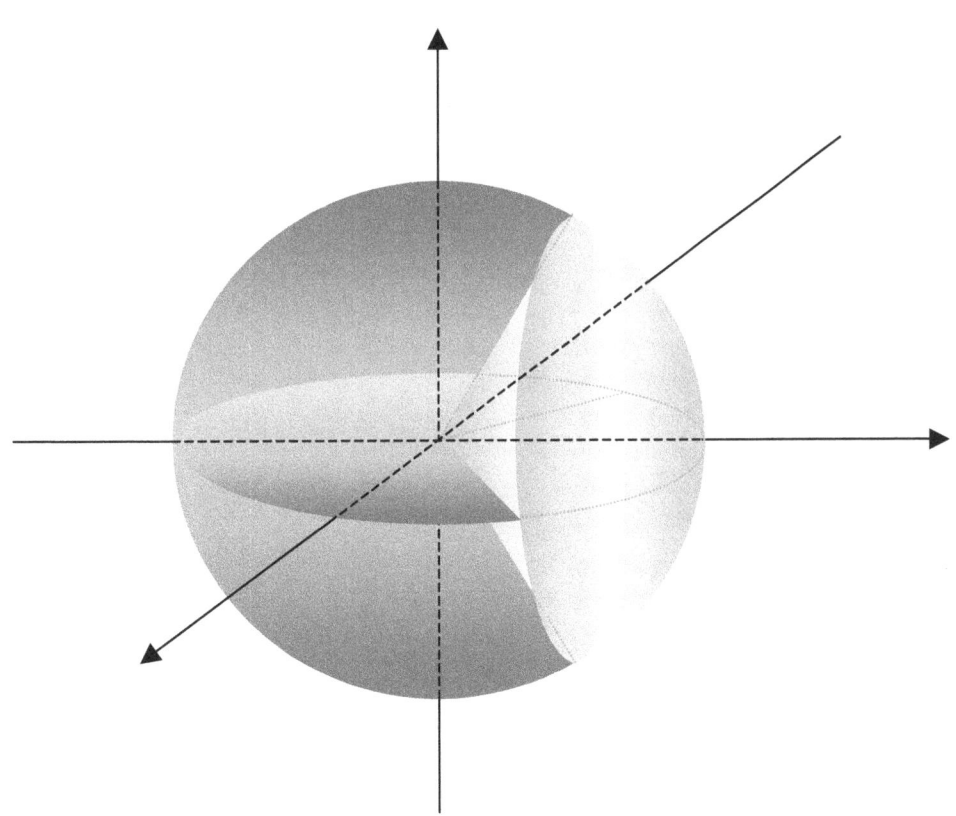

Seong R. Kim

Dear students:

Students need the best teacher, so you need examples, because examples are the best teacher. All the examples here are fully worked, and explain **how** the basic and essential tools in math are made, together with **what** they are, **how** they work, and **how** to work with them. Such tools include numbers, formulas, identities, equations, laws, etc.

Examples here begin with easy ones, of course. Covering every meter and yard properly, we can cover thousands of miles and kilometers. And it is particularly the case in math.

Of those examples therefore, some might even look too easy for you. It's not that easy though, to come up with those examples. Anyways, the bigger and the taller the tree, the deeper and the stronger the root.

Doing math, we work with ideas and run ideas, because every thing in math is an idea. A number is an idea, for instance, and the same is true for a line or circle, too. And putting ideas together, we build another, which becomes the base or an element of another, and each is connected. And that's the way your math grows. So you get to build a circuit, and sometimes, need to fill the gap or repair the circuit so that you get the sense of it.

So your calculation runs properly, and you get the problem solved.

The examples have been made and arranged so that they get tougher (or sometimes easier for some reason) as you proceed with them. In particular, similar examples with some variations are strategically repeated so that you can get the ideas or the tools tricky or complicated, and can get them mastered.

This book is however, nothing but a bunch of examples until you get it powered. How then, to get it powered, and make it run and work for you?

Just read it, and then, do each example in writing. And it is important to note that you do it in **your** writing. Just watching someone doing it, you just only feel that you can do it. If you do it, you can do it, but if you don't, we can hardly. It's a cliché, of course, but is always true that knowing is one thing and doing is another.

I've been helping students grow, take care of, and run their own math. The area covers algebra and geometry for high school or college students, and is especially for equations (for unknowns or curves), functions, and their graphs, which are the basic elements in calculus, which's been the core of my interest from my early age in high school.

Of my students, some are quite poor in math, and thus, are afraid of or hate math, some require special education because of exceptional intelligence, some are smart enough, some are naïve and diligent, some are clever but lazy, and most behave in general. All the students are badly after though, one thing in common: a strong and secure math skill. It is of course, the prime objective of my work, and I'm always happy to and eager to help them achieve it. The problem was however, that many of them wanted it to be purchased. And the question is, can we buy it?

We can buy the means, of course. And a solid math skill is feasible, too. We know however, we can't buy love, and the same is true for the math skill, too. It's not what we can buy or sell, and not what we can give or take. It is however, what we can grow, and need to grow. Your math grows as much as you grow and take care of it. So does mine.

What math then, do students most often do or use in high schools or colleges?

It is algebra and geometry. What algebra though?

Elementary algebra, of course
Doing the algebra, we work with numbers (many in kinds), constants, variables, ratios, rates, expressions, equations, inequalities, functions, identities, formulas, laws, etc., together with signs and symbols. And if we want to do algebra properly, we want to know their natures and how they mingle with each other.

So studying math ideas or tools, you want to know **what** they are, **how** they work, and **how** to work with them or **what** to do with them.   What then, about the geometry?

Basically, the geometry has much to do with shapes, positions, and angles. The shapes begin with triangles and circles, and move on to rectangles, squares, parallelograms or rhombuses, trapezoids, tetragons, other polygons, polyhedrons, etc.

Doing the geometry, too, though, we need to do the algebra stated above. So it is analytic geometry, often called coordinate geometry, too. And doing it, we can specify positions using coordinates. So in the geometry, basically, we work with graphs. Putting a math idea in a graph, we can not only effectively think about it but actually see it, too, and therefore, can efficiently work with it. What idea then, is it?

The idea begins with a point, line, parabola, circle, ellipse, and hyperbola, called a conic section or basic curve, and then, moves on to other curves, planes, surfaces, volumes, and other objects in various dimensional spaces, together with vectors.

And using an angle, we can specify an amount of turn or change in direction.

So learning, using, or applying those ideas or math tools, we get to solve problems.

And this book can help. It can help learn them, and use them so that you can navigate to find solutions to problems. And in particular, it can help come up with answers to those **what**s and **how**s stated above. So it can help you grow and run your own math, and thus, can help achieve your solid math skill.

It is however, not a magic book giving you a math skill of high caliber overnight. And it can have many mistakes, too. There is no magic, and math is full of facts and ideas. And it is after all, not me and not your teacher but you who put together some of those facts and ideas, and understand it. Putting facts and ideas together, understanding it, and taking care of what you have learned, you grow your math. And this book can help.

This is a book of examples designed to help you grow your math, and assumes that you are a real beginner. This book requires though, time and effort, the amount of which need to be substantial, too, but will be worth it. That's because you want a substantial achievement, and will get it. And probably, you will get to see this book helping you get there much faster than expected. And then, you will get to see the way math runs.

In math, everything is an idea. So is a problem. And solving it, we put it many different ways. For instance, while expanding or reducing it, or modifying or converting it, we keep searching for the solution, approaching the solution, and eventually, can get there. So don't look for the solution outside the problem. The solution is inside the problem if the problem is properly made.

If it is not, no solution is the solution. And in fact, it is often the case a problem itself is the solution. We can put a problem in many different ways, and eventually, can end up with the solution.   How come then, is the solution no other than the problem?

For instance, the solution to $3232 \div 101$ is 32.   And we can put it this way:

$$3232 \div 101 = \frac{3232}{101} = \frac{32 \times 101}{101} = \frac{32}{1} = 32 \Rightarrow 3232 \div 101 = 32.$$

And we can get this, too: $32 \Rightarrow 3232 \div 101$.   How?

$$32 = \frac{32}{1} = \frac{32 \times 101}{101} = \frac{3232}{101} = 3232/101 = 3232 \div 101.$$ Too easy?

For another instance, the solution to $ax^2 + bx + c = 0$ is: $x = \frac{-b \pm \sqrt{b^2 - 4ac}}{2a}$, which is called the quadratic formula.   How come then, is the solution no other than the problem?

We can put it this way:

$$x = \frac{-b \pm \sqrt{b^2-4ac}}{2a} \Rightarrow 2ax = -b \pm \sqrt{b^2 - 4ac} \Rightarrow 2ax + b = \pm\sqrt{b^2 - 4ac}$$

$$\Rightarrow (2ax + b)^2 = b^2 - 4ac \Rightarrow 4a^2x^2 + 4abx + b^2 = b^2 - 4ac$$

$$\Rightarrow 4a^2x^2 + 4abx = -4ac \Rightarrow ax^2 + bx = -c \Rightarrow ax^2 + bx + c = 0.$$

And we can get this, too: $ax^2 + bx + c = 0 \Rightarrow x = \frac{-b \pm \sqrt{b^2-4ac}}{2a}$.   How?

$$ax^2 + bx + c = a(x^2 + \tfrac{b}{a}x) + c = a(x^2 + \tfrac{b}{a}x + \tfrac{b^2}{4a^2} - \tfrac{b^2}{4a^2}) + c = a(x^2 + \tfrac{b}{a}x + \tfrac{b^2}{4a^2}) - \tfrac{b^2}{4a} + c$$

$$= a(x + \tfrac{b}{2a})^2 - \tfrac{b^2-4ac}{4a} = 0 \Rightarrow a(x + \tfrac{b}{2a})^2 = \tfrac{b^2-4ac}{4a} \Rightarrow (x + \tfrac{b}{2a})^2 = \tfrac{b^2-4ac}{4a^2} \Rightarrow x + \tfrac{b}{2a} = \pm\sqrt{\tfrac{b^2-4ac}{4a^2}}$$

$$\Rightarrow x = -\tfrac{b}{2a} \pm \tfrac{\sqrt{b^2-4ac}}{2a} = \tfrac{-b \pm \sqrt{b^2-4ac}}{2a} \Rightarrow x = \tfrac{-b \pm \sqrt{b^2-4ac}}{2a}.$$

And we call the set of processes above, algebra.

So if a problem is well defined, that is, if it makes sense, we should be able to get it solved the way below:

A problem ⇒ … ⇒ … ⇒ the solution, and thus: **the problem ⇒ the solution**.

So solving a problem, we put it many different ways so that we can get to the solution.

And that's the way, math runs.

May your math run very well.

Seong R. Kim

B.S. Math. Michigan Tech. Univ.   M.S. Math. Rensselaer Polytechnic Institute

Notes:

This book is about an idea called functions. Why functions, though?

Using functions, we can see how things change.
More specifically, how values change as other values change.   How come?

Basically, functions are about values that change.
Expressing values, we use numbers, of course.
Not only that, of course. But we use letters, too, called variables or constants.
And using those letters, along with numbers, we make expressions that express values.
So values of expressions change as values of variables change.
And we want to know how their values change as values of their variables change.

And we call such an expression a function. So we want to know how the value of a
function changes as the value of the variable changes.
And thus, doing physics, economics, or anything that have to do with changes, we want
to know about functions. In short, how things change.

Increase or decrease
Going up or coming down
Speeding up or slowing down

So how fast or slow is it? And by how much or how many?

For instance:
How the distance changes as time changes.
How the pressure changes as temperature changes.
How the price changes as supply decreases or demand increases.

And after all, how outputs change as inputs change.
Functions get inputs as time, and in return, produce outputs as distance.

So if need to work with things change, you want to know functions, and how they work, together with how to work with them.

And functions are expressions mainly made of polynomials, made of variables, constants, and numbers, of course. So you want to know polynomials, how they work, and how to work with them, too. And as stated above, polynomials are expressions made of variables, constants, and numbers, of course. You want to know thus, variables and constants, how they work, and how to work with them, along with their arithmetic.

And this book covers all the basics and ideas stated above. What then, about numbers?

A function can use numbers of all kinds. Or we can say that we can use in a function number of all kinds. Mainly though, we use numbers said to be real. So at least, you want to know real numbers of all kinds, and we often put them into three groups: integers, rational numbers, and irrational numbers.

And we have numbers that have different looks. Among those, we often use powers, and have logarithms, called logs, too. And that's not it. We have radicals, called roots, too. And using those above, we can make expressions called functions. So you want to cover quite a few to learn, make, and use functions. So this book covers the basics and ideas using all such numbers when working with functions.

And we can actually see how functions behave, too. Each function can have its own look, which is called its curve, called a graph, too. And producing the graph or curve of a function, we can say we put the function in a graph or just we graph it. So putting the a function in a graph or graphing a function, you can see how the function behaves, and can find readily and fast the one you are busy looking for. And the one is of course, the solution to the problem you are busy doing. So this book shows many examples of graphs of many kinds.

Besides, though this book, you can improve your skill of algebra, too.
So the book does not just explain the things stated above. But it also helps follow steps to the solutions, and thus, helps you do calculations with expressions so that you can improve your calculation work when working with functions.

With strong skill of algebra, you can do a lot, and of course, can do problems very well, too. And through this book, you can grow much of your power in algebra.

And all the basics and many ideas stated above are covered in two separate books. And the two books are as follows:

ALGEBRA EXAMPLES BASIC FUNCTIONS 1

ALGEBRA EXAMPLES BASIC FUNCTIONS 2

And all the contents in the two books above are put together in one book as follows:

ALGEBRA EXAMPLES BASIC FUNCTIONS

So this book will help you learn what functions are about, how they work, and what to do with them, together with how to do it. And also, this will help improve your skill of algebra, too. You will soon be thus, able to change or alter, convert, or modify math expressions so that you can get to the solutions fast. And you will learn them through examples detailed so that your math can run not only properly but fast enough, too.

Contents

In BASIC FUNCTIONS 2

The Preview of the Contents

In BASIC FUNCTIONS 1

Note:

The drawings or graphs in this book are not exact, and are approximate or conceptual ones.

$\in$	"$a \in B$" means that a belongs to B. "$p, q,$ and $r \in W$" means that $p, q,$ and r belong to W.
$\Rightarrow$	"$A \Rightarrow B$." means that A implies B.
$\equiv$	$A \equiv B$ means that A and B are identical to each other.
$\neq$	$A \neq B$ means that A is not equal to B.
$|A|$	The magnitude of A. For instance, $|\text{-}1| = |1| = 1$.
$\therefore$	Therefore
$\Leftrightarrow$	"$A \Leftrightarrow B$" means "If A then B." and "If B then A." We can read $A \Leftrightarrow B$ as "A if and only if B." In such a case, we can say that $A = B$.
Δx and Δy	Suppose that (x_1, y_1) and (x_2, y_2) are two points in the x-y plane. Then, we get either of the two below. $\Delta x = x_2 - x_1$, and $\Delta y = y_2 - y_1$. $\Delta x = x_1 - x_2$, and $\Delta y = y_1 - y_2$.

Distance Formula

Suppose that d is the distance between two points (x_1, y_1) and (x_2, y_2) in the x-y plane. Then, we get $d^2 = (\Delta x)^2 + (\Delta y)^2$.

5.0. Inverse Functions 1

Why is it inverse, though?

As mentioned earlier on one-to-one functions, we sometimes need to come up with a new function from a function given.

Coming up with such a function, we need to use as the domain the range of the function given, and use as the range the domain of the given function.

So a domain becomes a range, and a range becomes a domain. It's just one thing though.

We need to keep the existing correspondence between numbers in the domain and range of the function given. That is, the number pairs have to remain. Only roles are reversed.

For instance, if the given function makes (1, 2), (4, 7), (6, -5) etc., the new function has to make (2, 1), (7, 4), (-5, 6), etc.

We call such a new function an inverse function.
And taking the inverse function of a function given, we call it the inverse of the given function, and often just call it *the inverse*, for short.

Suppose for instance, that we are given a function f of which the domain is X, and the range is Y, and that f is one-to-one.

Then, we can come up with a new function of which the domain is Y and the range is X. And the new function will have a new expression other than that of f.

That's because the new expression has to maintain the existing number pairs, but the two numbers in each pair have to exchange their positions.

So for instance, assuming **g** is the new function, and *f*(4) = 7, we need to get: **g(7) = 4**.

And thus, the dependence relationship is inverted. So we call such a new function the inverse of the function given, and just call it the inverse, too, for short.

Thus, **g** is called the inverse of *f*, and we say that such a function as *f* is invertible.

So if a function is invertible, it is said to have its inverse, and if not invertible, it is said to have no inverse.

Finding or making the inverse of a function, we say we take the inverse of the function. And a function has to be, of course, invertible if we want to take the inverse of the function. What function then, can be invertible?

If it is invertible, it is one-to-one, and if one-to-one, it is invertible.
So only functions one-to-one can be invertible. And we should be able to get the inverse of a function one-to-one. Why one-to-one, though?

Of a function, <u>each input has to get paired with one output only</u>, which is a rule of a function. That is, each element in a domain gets paired with one element only in the range, and thus, is *not* allowed to get paired with another element in the range.

So for instance, assuming **h(1) = 3**, we cannot get: **h(1) = 5** or any number other than 3.

Fig. 0   Domain   Range

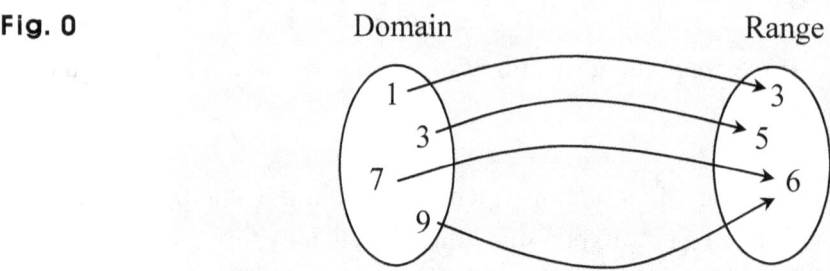

Let's this time, look at both sets the other way around.

Fig. 1   Domain   Range

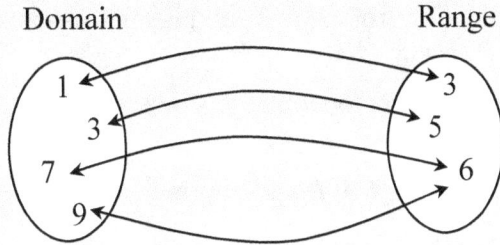

Then, unlike an input, we can see that an output can get paired with more than one input, which is in fact, another rule of a function.

That is to say that each element in the range can get paired with more than one element in the domain.
So for instance, even though $h(7) = 6$, we can get: $h(9) = 6$, too.
That is to say that a function can make number pairs the way as follows: (7, 6) and (9, 6).

And thus, an element in the range gets paired with an element in the domain, and yet, can still get paired with another element or even more in the domain.   So what?

- If a function has the inverse, the function has to be one-to-one.

Otherwise, the inverse would not be a function. If the function we take the inverse of is many-to-one, the inverse would be one-to-many, but no function is one-to-many.

Suppose for instance, k were the inverse of h above.

Then, we would get not only $k(6) = 7$ but $k(6) = 9$, too, which is however, not allowed. No function makes number pairs this way: (6, 7) and (6, 9). So k cannot be a function.

That is to say that h cannot have the inverse.

That is because of the fact that h is not one-to-one.

So only functions one-to-one can be invertible, and such a function has the inverse.

How then, can we get the inverse function?

Suppose again, f is a function from X to Y, and is one-to-one, and g is the inverse of f.

Then, of course, g is one-to-one, too, since f is one-to-one.

And the domain of g is the range of f, and the range of g is the domain of f.

That is, the roles of the data sets get reversed. So Y is the domain of g, and X is the range of g, which is thus, a function from Y to X. That is only one thing, though.

What's really about the inverse is the expression of the inverse.

Let's now call the function f the original function.

Then, what matters is how the inverse produces all the inputs of the original function using all the corresponding outputs of the original.

So what we want to know is the way the input changes as the output changes. How then, can we find the way?

Assuming for instance, $y = f(x) = x + 1$, we can say that y changes as x changes. How then, does y change? In other words, what is the way y changes as x changes?

It is: $x + 1$. It is saying that for each and every value of x, the y-value is the sum of 1 and the x-value. So the expression of f shows the way y changes as x changes.

That is to say that the expression of f shows the way the output changes as the input changes. What then, do we need to get finding the inverse of f?

Finding the inverse, we want to find the way the input changes as the output changes. And the way is the expression of the inverse. How then, can we find the expression?

We don't just find it, of course. We can try finding it using the expression of the original.

We have: $y = f(x) = x + 1$, so we get: $y = x + 1$.

Then, we can say that y is set equal to an expression expressed in terms of x.
In short, y is an expression in terms of x.   What then, do we mean by the expression?

The expression shows how the y-value gets made for each and every x-value.
What then, do we need to get to find the way the x-value gets made for each and every y-value?

We want to get the equation where x is set equal to an expression expressed in terms of y.
In short, we need an expression in terms of y.

It's because the expression shows how the x-value gets made for each and every y-value.
How then, can we get the expression?

The expression of the inverse is normally different from that of the original.
In what case then, the expression of the inverse is the same as that of the original?

If the original is an identity function, the expression of the inverse is that of the original, which is, of course, quite trivial. So it's not worth a talk.   What if the original is not an identity function?   How then, can we get the expression of the inverse?

The expression shows the way the x-value gets made as the y-value changes.   So?

The expression of the original is the way the y-value gets made as the x-value changes.
That is to say that the expression of the original connects the y-value and the x-value.

So we can try finding the expression of the inverse using the expression of the original.
How though?

We have: $y = f(x)$, which can be taken as an equation for x, because $f(x)$ is an expression in terms of x as $2x$, $x^2 - 1$, $x + 3$, etc. Then, we have: $y = 2x$, $y = x^2 - 1$, $y = x + 3$, etc.

What then, do we get solving for x, the equation $y = f(x)$?

We can get an equation, where the input variable x is set equal to an expression expressed in terms of y, that is, the output variable.

So the expression show the way the x-value gets made as the y-value changes.
And the expression is the very expression of the inverse.

We need to note however, that we normally put a function in the x-y system.

Putting a function in the x-y system, we take x as the input variable, and take y as the output variable.

So once we've got the equation stated above, we need to swap the two variables.
That is, we want to replace x with y, and replace y with x.

Then, we produce a function definition that describes the inverse.
A function definition is a description of a function as $y = p(x) = x^2 + 3x + 1$ for $x > 2$.

Let's now find the inverse of the function f assuming the inverse is g.

We have: $y = f(x) = x + 1$. So solving for x, the equation $y = x + 1$, we get: $x = y - 1$, which is an expression in terms of y, and is the very expression of the inverse g.

- So *for now*, we can put g the way as follows: $x = g(y) = y - 1$.

Then, in the inverse function g above, y gets inputs, and thus, is the input variable, and x gets outputs, and thus, is the output variable.

By convention though, we normally take x as an input variable, and take y as an output variable. That is, we normally put a function in the x-y system.

- Producing thus, the inverse g, we want to put it this way: $y = g(x) = x - 1$.

Then, of course, the inverse function g produces all the inputs of the original function f using as inputs, all the corresponding outputs of the original f.

Suppose now, for another simple instance, of f, each output is twice each input.

Then, of the inverse g, each output has to be half each input.
So the expression of $f(x)$ is: $2x$, and thus, the expression of $g(x)$ is: $\frac{1}{2}x$.

Solving in fact, for x first, the equation $y = 2x$, we get: $x = \frac{1}{2}y$.

And next, swapping x and y, we get: $y = \frac{1}{2}x$, which is the expression of the inverse

function $y = g(x)$. That is, the inverse is: $y = g(x) = \frac{1}{2}x$.

Let's find, for another instance, the inverse of $y = f(x) = 2x + 1$.

Then, solving for x first, the equation $y = 2x = 1$, we get: $2x = 1 - y \Rightarrow x = (1 - y)/2$.
Then, swapping the variables, we get: $y = (1 - x)/2$.
Assuming thus, the inverse is g, we get: $y = g(x) = (1 - x)/2$.

What then, about this case: $y = f(x) = \sqrt{x + 4} + 1$ for $x > 5$?

Solving for x, the equation $y = f(x)$, we get: $x = y^2 - 2y - 3$.
Then, swapping x and y, we get: $y = x^2 - 2x - 3$, which is the very expression of the inverse function $y = g(x)$.

We don't just produce however, the definition of the inverse the way as follows:
$y = g(x) = x^2 - 2x - 3$.   Why not?

That's because the inverse is a function one-to-one.
And if not specified, the domain is assumed to be the largest set of numbers that can be the values of the input variable. So in the case above, the domain is assumed to be a set of all real numbers. And thus, the function g produced the way above is not one-to-one.

So we need to find and specify the domain of the inverse.

Finding in fact, the range of f, we get: $y \geq 4$.   So the domain of the inverse is: $x \geq 4$.

Thus, the inverse is: $y = g(x) = x^2 - 2x - 3$ for $x \geq 4$, which is one-to-one.

What then, about this case: $y = f(x) = x^2 - 2x - 3$ for $x > 3$?

We've got to solve if for x, too, and then, swap the variables in the equation we get as the solution. That's not it though.
We need to specify the domain of the inverse, and determine its expression considering the domain, too.

So in reality, what really matters is your caliber in algebra.
You need to have a strong foundation on algebra. What algebra?

Elementary algebra, together with geometry analytic, called coordinate geometry, too.

And the algebra needs to include polynomial factorizations, which covers how to manipulate expressions, that is, how to change, alter, or convert expressions so that you can actually get to the very solution you need to put down on your paper.

Examples 1 in Inverse Functions

Though these examples are for your skill on functions, particularly on inverse functions, they are for your algebra skill, too. And the same is true, too, for all the other examples in this book.

Math skill is about problem solving skill, that is, the skill to get solutions to problems. And what actually connects problems to solutions is algebra. What algebra?

Doing algebra, you get to manipulate expressions, that is, you need to change or alter, convert or transform expressions the way you can get the ones you are looking for. What expressions?

Most often used are numbers of many kinds, variables and constants, monomials and polynomials, powers and logarithms, radicals and fractionals, ratios and rates, and equations and functions.

And what matters is *your* algebra, that is, *your* calculation skill, and more specifically, how to handle those expressions to get to the ones you are busy looking for. And you know what you are busy looking for doing problems. They are the solutions, of course.

0.   Find the inverse of $y = f(x) = ax + b$ where a and b are constant, and $a \neq 0$.

1. Find the inverses of the functions as follows.

0. $y = f(x) = \sqrt{2 - x} + 3$ for $x \leq 2$.

1. $y = f(x) = (2x - 1)(2x + 5)$ for $x \geq 0$.

Suggestions or Solutions
To the **Problem** in the Example **0**

Find the inverse of $y = f(x) = ax + b$ where a and b are constant, and $a \neq 0$.

$y = ax + b \Rightarrow x = \frac{1}{a}(y - b).$

So if **g** is the inverse, we get: $y = g(x) = \frac{1}{a}(x - b).$

If not quite sure of the idea behind the processes above, follow the steps below:

Finding the inverse of a function, what do we need to begin with?

We may want to begin with checking to see if the function is one-to-one.
Usually though, if an inverse is asked, the original function is assumed to be one-to-one.
Anyway, the function f is one-to-one.   How come?

The curve of f is a line, so no output can appear more than once.

Checking analytically though, we can do either a division or subtraction with two arbitrary outputs as $f(u)$ and $f(v)$ where u and v are constant, and $u \neq v$, of course.

And if the quotient is 1 or the difference is 0, the function f is not one-to-one.

Otherwise, it is one-to-one. So let's try the testing for practice.

Taking the difference, we get: $f(u) - f(v) = (au + b) - (av + b) = a(u - v) \neq 0.$

So $f(u) \neq f(v)$, and thus, f is one-to-one.

What then, is the next to do?

We want to find the expression of the inverse.   How?

We want to solve for the input variable the equation made of the expression of the original function, then swap the variables.
So first, we want to solve for x the equation $y = ax + b$.

Then, we get: $y = ax + b \Rightarrow x = \frac{1}{a}(y - b)$. And next, swapping the variables, we get: $y = \frac{1}{a}(x - b),$ which is the expression of the inverse.

So assuming g is the inverse, we get: $y = g(x) = \frac{1}{a}(x - b)$ for x real.

How come in the inverse, the domain is a set of all real numbers?

Of the original function f, the range is a set of all real numbers, and the range is the domain of the inverse. And in fact, the domain of f is assumed to be a set of all real numbers since the domain is not specified.

In short:

$y = ax + b \Rightarrow x = \frac{1}{a}(y - b).$

So assuming g is the inverse, we get: $y = g(x) = \frac{1}{a}(x - b)$ for x real.

Suggestions or Solutions
To the **Problem 0** in the Example **1**

Find the inverse of $y = f(x) = \sqrt{2-x} + 3$ for $x \leq 2$.

$$y = \sqrt{2-x} + 3 \Rightarrow \sqrt{2-x} = y - 3 \Rightarrow 2 - x = (y-3)^2 \Rightarrow x = 2 - (y-3)^2.$$

The domain of f is: $x \leq 2$. So finding the range, we get:

$$x \leq 2 \Rightarrow \sqrt{2-x} \geq 0 \Rightarrow \sqrt{2-x} + 3 \geq 3. \quad \text{And thus, the range is: } y \geq 3.$$

So the domain of the inverse is: $x \geq 3$.

And thus, assuming g is the inverse, we get: $y = g(x) = 2 - (x-3)^2$ for $x \geq 3$.

If not quite sure of the idea behind the processes above, follow the steps below:

The function f is one-to-one. If not sure though, you may want to give it a check.

Now, what do we need to begin with finding the inverse?

Since f is one-to-one, we should be able to find the inverse. Finding the expression first, we want to solve for x the equation made of the expression of f, then swap the variables.

So solving the equation for x, we get:

$$y = \sqrt{2-x} + 3 \Rightarrow \sqrt{2-x} = y - 3 \Rightarrow 2 - x = (y-3)^2 \Rightarrow x = 2 - (y-3)^2.$$

And next, swapping the variables, we get: $y = 2 - (x-3)^2$.

Why swapping though?

The inputs of the inverse are the outputs of the original function, and the outputs of the inverse are the inputs of the original.

Now, is that all?

Not quite, and in fact, a critical factor is missing, and is the domain of the inverse.

What then, is the domain?

The domain is the range of the original function f, so we want to find the range of f.

The domain of f is: $x \leq 2$. So finding the range, we get:

$x \leq 2 \Rightarrow \sqrt{2-x} \geq 0 \Rightarrow \sqrt{2-x} + 3 \geq 3$. And thus, the range is: $y \geq 3$.

So the domain of the inverse is: $x \geq 3$.

And thus, assuming g is the inverse, we get: $y = g(x) = 2 - (x-3)^2$ for $x \geq 3$.

What if we use the original function name indicating the inverse?

(Using f^{-1} instead of g, we indicate the inverse, too.)

The definition for inverse functions is: $y = f(x) \Leftrightarrow x = f^{-1}(y)$.

And we know the inverse function $x = f^{-1}(y)$ is a function of y.

And also, we have its expression in terms of y, which is: $2 - (y-3)^2$, found above.

So we can put the inverse this way, too: $x = f^{-1}(y) = 2 - (y-3)^2$ for $y \geq 3$.

Normally though, we put functions in the x-y system where x is the input variable.

14

So we usually use x as the input variable, along with another function name as g. And of course, in such a case, we use y as the output variable in g.

In short:

$$y = \sqrt{2-x} + 3 \Rightarrow \sqrt{2-x} = y - 3 \Rightarrow 2 - x = (y-3)^2 \Rightarrow x = 2 - (y-3)^2.$$

The domain of f is: $x \leq 2$. So finding the range, we get:

$$x \leq 2 \Rightarrow \sqrt{2-x} \geq 0 \Rightarrow \sqrt{2-x} + 3 \geq 3. \quad \text{And thus, the range is: } y \geq 3.$$

So the domain of the inverse is: $x \geq 3$.

And thus, assuming g is the inverse, we get: $y = g(x) = 2 - (x-3)^2$ for $x \geq 3$.

Let's now put the curves of both functions, and see how they behave.

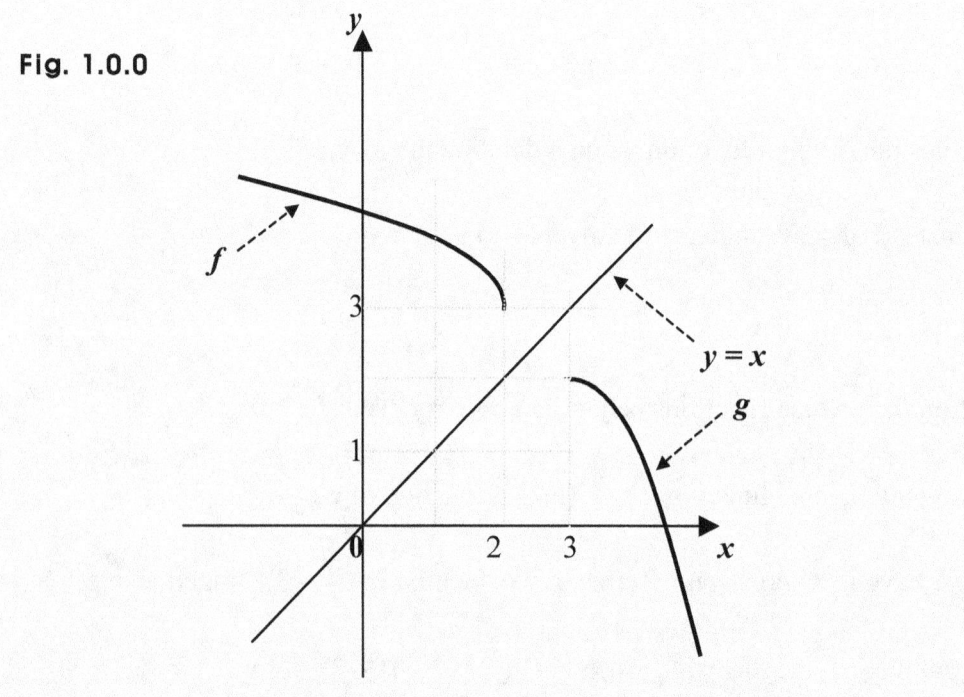

Fig. 1.0.0

Both curves are symmetric about the line $y = x$.

Suggestions or Solutions
To the **Problem 1** in the Example **1**

Find the inverse of $y = f(x) = (2x - 1)(2x + 5)$ for $x \geq 0$.

$$y = (2x-1)(2x+5) = 4x^2 + 8x - 5 = 4(x^2 + 2x) - 5 = 4(x^2 + 2x + 1 - 1) - 5$$

$$= 4(x^2 + 2x + 1) - 4 - 5 = 4(x+1)^2 - 9 \Rightarrow y + 9 = 4(x+1)^2$$

$$\Rightarrow (x+1)^2 = \tfrac{1}{4}(y+9) \Rightarrow x + 1 = \pm\tfrac{1}{2}\sqrt{y+9} \Rightarrow x = \pm\tfrac{1}{2}\sqrt{y+9} - 1.$$

$$x \geq 0 \Rightarrow (x+1)^2 \geq 1 \Rightarrow 4(x+1)^2 \geq 4 \Rightarrow 4(x+1)^2 - 9 \geq \text{-}5.$$

So the range of f is: $y \geq \text{-}5$. So using the range to check the domain, we get:

$$y \geq \text{-}5 \Rightarrow \sqrt{y+9} \geq 2 \Rightarrow \tfrac{1}{2}\sqrt{y+9} \geq 1 \Rightarrow \tfrac{1}{2}\sqrt{y+9} - 1 \geq 0 \Rightarrow x \geq 0, \text{ the domain of } f.$$

$$y \geq \text{-}5 \Rightarrow \sqrt{y+9} \geq 2 \Rightarrow \text{-}\tfrac{1}{2}\sqrt{y+9} \leq \text{-}1 \Rightarrow \text{-}\tfrac{1}{2}\sqrt{y+9} - 1 \leq \text{-}2 \Rightarrow x \leq \text{-}2, \text{ not the domain.}$$

And thus, assuming g is the inverse, we get: $y = g(x) = \tfrac{1}{2}\sqrt{x+9} - 1$ for $x \geq \text{-}5$.

If not quite sure of the idea behind the processes above, follow the steps below:

Normally, if the domain is a set of all real numbers, and the expression is a polynomial of even degree as quadratic, the function is not one-to-one.

In the given function f, the expression is quadratic polynomial, and yet, the domain is not the entire real number space but a set of all nonnegative numbers.

So the function f can be one-to-one. And thus, we may want to give a check.
In this case though, use of two arbitrary outputs is not quite useful because the algebra looks quite messy.   How then, can we see if it is one-to-one or not?

Putting the curve in a graph, we can quickly see if it is one-to-one.

The curve of f is a parabola, so using the sign of the coefficient of the quadratic term and the axis of symmetry, we can readily see if it is the case, too.

So let's now check to see if it is one-to-one.

To begin with, expanding (simplifying) the expression of f, we get:
$(2x - 1)(2x + 5) = 4x^2 + 8x - 5$.

Thus, the coefficient is 4, which is positive, so the parabola is concave-up.
And the axis of symmetry is a line: $x = -1$.   How come?

If a parabola is: $y = ax^2 + bx + c$, the axis of symmetry is: $x = -\frac{b}{2a}$, which is a line parallel to the y-axis. And we have: $y = 4x^2 + 8x - 5$, so $a = 4 > 0$, and the axis is: $x = -1$.

Now, putting threads together, the coefficient is positive, and the axis of symmetry is outside the domain of f, which is: $x \geq 0$. So the function f is one-to-one.

(If not sure, refer to **ALGEBRA EXAMPLES CONICS 2**.)

Now, what then, is the next to find the inverse?

Since f is one-to-one, we should be able to find the inverse. Finding the expression first, we want to solve for x the equation made of the expression of f, then swap the variables.

So solving the equation for x, we get:

$y = 4x^2 + 8x - 5 = 4(x^2 + 2x) - 5 = 4(x^2 + 2x + 1 - 1) - 5 = 4(x^2 + 2x + 1) - 4 - 5$

$= 4(x + 1)^2 - 9 \Rightarrow y + 9 = 4(x + 1)^2$

$\Rightarrow (x + 1)^2 = \frac{1}{4}(y + 9) \Rightarrow x + 1 = \pm\frac{1}{2}\sqrt{y + 9} \Rightarrow x = \pm\frac{1}{2}\sqrt{y + 9} - 1.$

And next, swapping the variables, we get: $y = \pm\frac{1}{2}\sqrt{x+9} - 1$. So we have two, one is: $y = \frac{1}{2}\sqrt{x+9} - 1$, and the other is: $y = -\frac{1}{2}\sqrt{x+9} - 1$. How come two, though?

An inverse has, of course, one expression. So we get to choose one of the two. How?

Using the range of the original function, we can make a right choice. How come?

Before swapping variables, we had: $x = \pm\frac{1}{2}\sqrt{y+9} - 1$.

And in the expression above, the extent of the values y can take is the range of f, and the extent of the values x can take is the domain of f.

We know the domain of f is: $x \geq 0$. And using the domain, we can find the range of f.

When solving the equation made of the expression of f, we have got: $y = 4(x + 1)^2 - 9$.

That is, we can set: $y = f(x) = 4(x + 1)^2 - 9$ for $x \geq 0$. So finding the range, we get:

$$x \geq 0 \Rightarrow (x + 1)^2 \geq 1 \Rightarrow 4(x + 1)^2 \geq 4 \Rightarrow 4(x + 1)^2 - 9 \geq \text{-}5.$$

And thus, the range of f is: $y \geq \text{-}5$. So applying the range to the two expressions, we get:

$y \geq \text{-}5 \Rightarrow \sqrt{y+9} \geq 2 \Rightarrow \frac{1}{2}\sqrt{y+9} \geq 1 \Rightarrow \frac{1}{2}\sqrt{y+9} - 1 \geq 0 \Rightarrow x \geq 0$, which is the domain of f.

$y \geq \text{-}5 \Rightarrow \sqrt{y+9} \geq 2 \Rightarrow -\frac{1}{2}\sqrt{y+9} \leq \text{-}1 \Rightarrow -\frac{1}{2}\sqrt{y+9} - 1 \leq \text{-}2 \Rightarrow x \leq \text{-}2$, which is not.

So the expression of the inverse is: $\frac{1}{2}\sqrt{y+9} - 1$.

And of course, putting it in the *x-y* system, we swap the variables.

So the expression of the inverse in the *x-y* system is: $\frac{1}{2}\sqrt{x+9}-1$. What then, is the next?

We want to get the domain of the inverse, which is the range of *f*, which has been though, already found, and is: $y \geq$ **-5**.

And thus, assuming **g** is the inverse, we get: $y = g(x) = \frac{1}{2}\sqrt{x+9}-1$ for $x \geq$ **-5**.

In short:

$$y = (2x - 1)(2x + 5) = 4x^2 + 8x - 5 = 4(x^2 + 2x) - 5 = 4(x^2 + 2x + 1 - 1) - 5$$

$$= 4(x^2 + 2x + 1) - 4 - 5 = 4(x + 1)^2 - 9 \Rightarrow y + 9 = 4(x + 1)^2$$

$$\Rightarrow (x + 1)^2 = \frac{1}{4}(y + 9) \Rightarrow x + 1 = \pm\frac{1}{2}\sqrt{y+9} \Rightarrow x = \pm\frac{1}{2}\sqrt{y+9}-1.$$

$$x \geq 0 \Rightarrow (x + 1)^2 \geq 1 \Rightarrow 4(x + 1)^2 \geq 4 \Rightarrow 4(x + 1)^2 - 9 \geq \text{-5}.$$

So the range of *f* is: $y \geq$ **-5**. So using the range to check the domain, we get:

$$y \geq \text{-5} \Rightarrow \sqrt{y+9} \geq 2 \Rightarrow \frac{1}{2}\sqrt{y+9} \geq 1 \Rightarrow \frac{1}{2}\sqrt{y+9}-1 \geq 0 \Rightarrow x \geq 0, \text{ which is the domain of } f.$$

$$y \geq \text{-5} \Rightarrow \sqrt{y+9} \geq 2 \Rightarrow -\frac{1}{2}\sqrt{y+9} \leq \text{-1} \Rightarrow -\frac{1}{2}\sqrt{y+9}-1 \leq \text{-2} \Rightarrow x \leq \text{-2}, \text{ which is not.}$$

And thus, assuming **g** is the inverse, we get: $y = g(x) = \frac{1}{2}\sqrt{x+9}-1$ for $x \geq$ **-5**.

And putting both curves in one graph, we get:

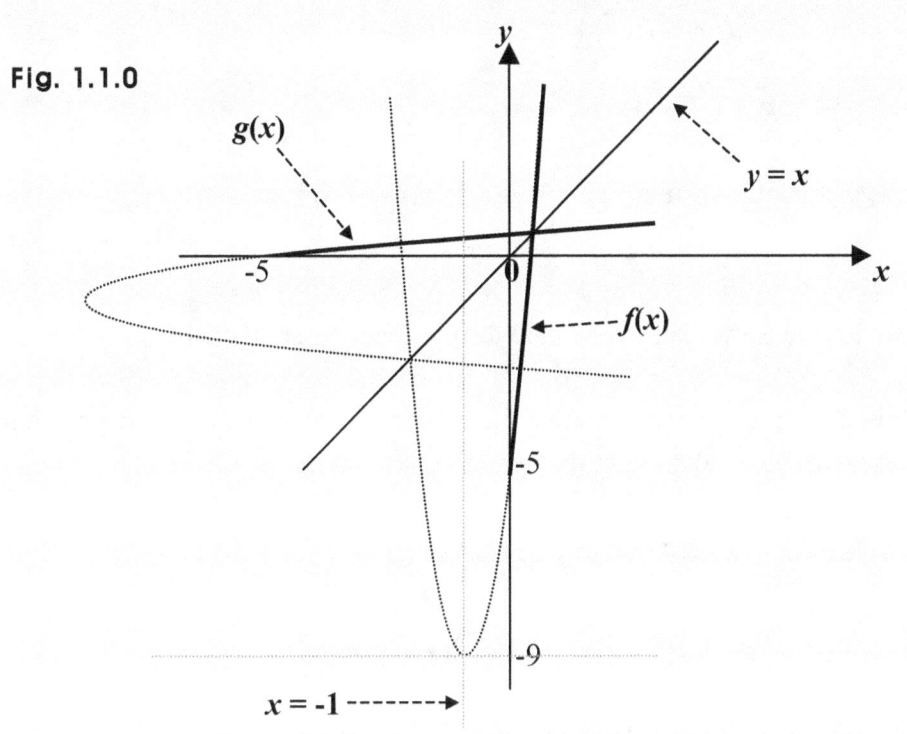

Fig. 1.1.0

And both curves are symmetric about the line $y = x$.

What if the function f is: $y = f(x) = (2x - 1)(2x + 5)$ for x real?

That is, the domain of f is the entire real number space, that is, a set of all real numbers.

Then, f is not one-to-one, since the same output appears twice as we can see it in the graph above. So can we not get the inverse?

Normally, we can't. We can find however, two functions, each of which is the inverse of f for a particular part of the domain that is a set of all real numbers.

So we want to partition the domain into two parts, one is: $x \geq -1$, and the other is: $x < -1$.

So to speak, we get to partition the function f into two functions, then find each inverse.

And if the two functions are f_1 and f_2, the two are as follows:

$y = f_1(x) = (2x - 1)(2x + 5)$ for $x \geq$ -1.

$y = f_2(x) = (2x - 1)(2x + 5)$ for $x <$ -1.

Then, each is one-to-one, so we should be able to find the inverse of each.

Finding the expression first, we want to solve for x the equation made of the expression of f, then swap the variables. So solving the equation for x, we get:

$y = (2x - 1)(2x + 5) = 4x^2 + 8x - 5 = 4(x^2 + 2x) - 5 = 4(x^2 + 2x + 1 - 1) - 5$

$= 4(x^2 + 2x + 1) - 4 - 5 = 4(x + 1)^2 - 9 \Rightarrow y + 9 = 4(x + 1)^2$

$\Rightarrow (x + 1)^2 = \frac{1}{4}(y + 9) \Rightarrow x + 1 = \pm\frac{1}{2}\sqrt{y + 9} \Rightarrow x = \pm\frac{1}{2}\sqrt{y + 9} - 1.$

And next, swapping the variables, we get: $y = \pm\frac{1}{2}\sqrt{x + 9} - 1.$
So we've got two expressions. Which one is which, though?

Using the range of each original function, we can make a right choice. How?

Before swapping variables, we had: $x = \pm\frac{1}{2}\sqrt{y + 9} - 1.$

And if we begin with f_1, in the expression above, the extent of the values y can take is the range of f_1, and the extent of the values x can take is the domain of f_1.

We know the domain of f_1 is: $x \geq 0$. And using the domain, we can find the range of f_1.

And the same is true for f_2, also.

When solving the equation made of the expression of f, we have got: $y = 4(x + 1)^2 - 9$.

That is, we can set: $y = f_1(x) = 4(x + 1)^2 - 9$ for $x \geq -1$.   So finding the range, we get:

$x \geq -1 \Rightarrow (x + 1)^2 \geq 0 \Rightarrow 4(x + 1)^2 \geq 0 \Rightarrow 4(x + 1)^2 - 9 \geq -9$.

And also, we can set: $y = f_2(x) = 4(x + 1)^2 - 9$ for $x < -1$.   So finding the range, we get:

$x < -1 \Rightarrow (x + 1)^2 > 0 \Rightarrow 4(x + 1)^2 > 0 \Rightarrow 4(x + 1)^2 - 9 > -9$.

And thus, the range of f_1 is: $y \geq -9$, and the range of f_2 is: $y > -9$.
So applying the ranges to the expressions of the inverses, we get:

$y \geq -9 \Rightarrow \sqrt{y+9} \geq 0 \Rightarrow \frac{1}{2}\sqrt{y+9} \geq 0 \Rightarrow \frac{1}{2}\sqrt{y+9} - 1 \geq -1 \Rightarrow x \geq -1$, the domain of f_1.

$y > -9 \Rightarrow \sqrt{y+9} > 0 \Rightarrow -\frac{1}{2}\sqrt{y+9} < 0 \Rightarrow -\frac{1}{2}\sqrt{y+9} - 1 < -1 \Rightarrow x < -1$, the domain of f_2.

So the expression of the inverse of f_1 is: $\frac{1}{2}\sqrt{y+9} - 1$.

And the expression of the inverse of f_2 is: $-\frac{1}{2}\sqrt{y+9} - 1$.

And of course, putting them in the x-y system, we swap the variables.

So the expression of the inverse of f_1 in the x-y system is: $\frac{1}{2}\sqrt{x+9} - 1$.

And the expression of the inverse of f_2 in the x-y system is: $-\frac{1}{2}\sqrt{x+9} - 1$.

And we know the range of f_1 is: $y \geq -9$, and the range of f_2 is: $y > -9$.

And thus, assuming g_1 is the inverse of f_1, we get: $y = g_1(x) = \frac{1}{2}\sqrt{x+9} - 1$ for $x \geq -9$.
And assuming g_2 is the inverse of f_2, we get: $y = g_2(x) = -\frac{1}{2}\sqrt{x+9} - 1$ for $x > -9$.
Putting both curves in one graph, we get:

22

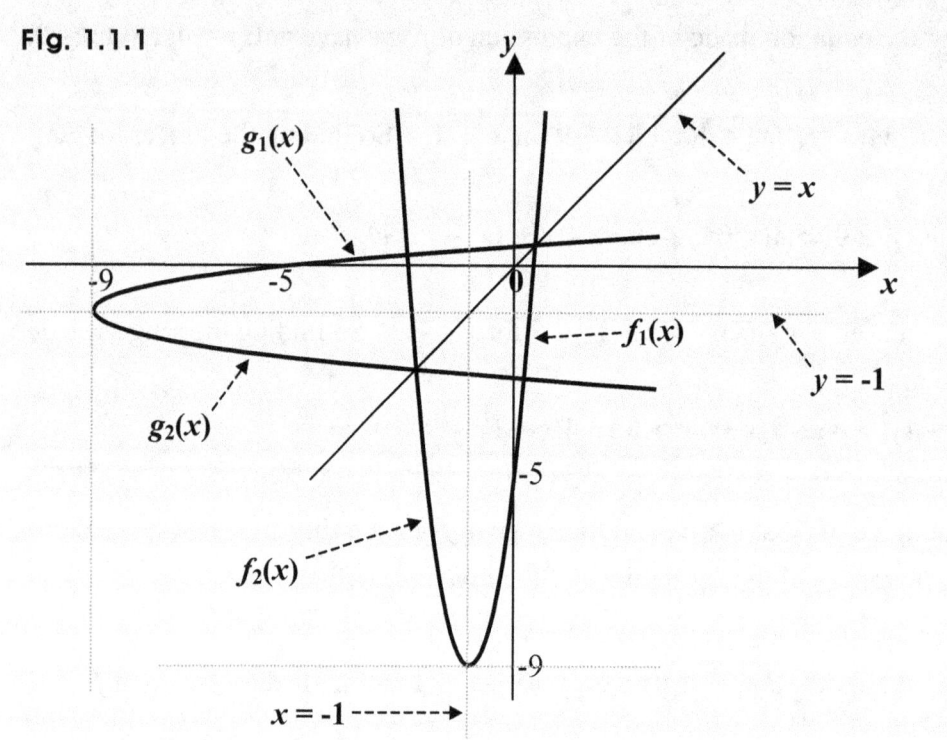

Fig. 1.1.1

Each curve is a half parabola.

The curve of g_1 is symmetric to the curve of f_1 about the line $y = x$.

And also, the curve of g_2 is symmetric to the curve of f_2 about the line $y = x$, too.

5.1. Inverse Functions 2

So what is an inverse function?

Answering the question, we produce <u>the definition for inverse functions</u>.
So the definition describes an inverse function, and says what it is.
What then, is the definition?

Assuming f is the original function, which is the function we take the inverse of, we can put the *definition for inverse functions* the way as follows:

X and Y are sets of real numbers.

$f: X \longrightarrow Y$, and f is one-to-one.

$g: Y \longrightarrow X, y = f(x) \Leftrightarrow x = f^{-1}(y)$, and $g = f^{-1}$.

Quite often though, we just use as the definition, the expression below:

$$y = f(x) \Leftrightarrow x = f^{-1}(y), \text{ and } g = f^{-1}.$$

And defining quickly a particular inverse, we can use a short definition, and can put it this way: $g = f^{-1}$, read as: g is the inverse of f. And f^{-1} is just read as: f inverse.

What then, can we do with the definition above?

We have some theorems on inverse functions, which can be of much help when we find inverse functions, and do problems with such functions. Understanding those theorems, we need to know the definition. And in particular, understanding inverse functions, we can better understand a log function, which is an inverse of an exponential function.

We are now going to cover some theorems on inverse functions. Math begins with definitions, so we may want to begin with the *definition for inverse functions* as follows.

X and Y are sets of real numbers.

$f: X \longrightarrow Y$, and f is one-to-one.

$g: Y \longrightarrow X, y = f(x) \Leftrightarrow x = f^{-1}(y)$, and $g = f^{-1}$.

Usually though, as the definition, we just use: $y = f(x) \Leftrightarrow x = f^{-1}(y)$.

And the theorems below can be of much help doing problems with inverse functions.

0. $(f^{-1})^{-1} = f$

1. $f^{-1}\{f(x)\} = x$, and in short: $f^{-1}(f) = x$.

2. $f\{f^{-1}(y)\} = y$, and in short: $f(f^{-1}) = y$.

3.   Assuming $y = f(x) = x$, we get: $f^{-1}(f) = f(f^{-1})$.

4.   If I_x is an identity function of x, and I_y is an identity function of y, we get:

$I_x = f^{-1} \bullet f, I_y = f \bullet f^{-1}$, and $(g \bullet f = I_x$ and $f \bullet g = I_y) \Leftrightarrow g = f^{-1}$.

5.   If $z = h(y)$ is one-to-one, we get: $(h \bullet f)^{-1} = f^{-1} \bullet h^{-1}$.

Let's see now, how the theorems can hold, and begin with the theorem 0: $(f^{-1})^{-1} = f$.

First, putting $(f^{-1})^{-1} = f$ more specifically, we get: $(f^{-1})^{-1}(x) = f(x)$.

And next, by the definition where $y = f(x) \Leftrightarrow x = f^{-1}(y)$, we get:

$y = f(x) \Leftrightarrow x = f^{-1}(y) \Leftrightarrow y = (f^{-1})^{-1}(x)$. Thus, $f(x) = (f^{-1})^{-1}(x)$.   What's going on?

Looking at the statement that $y = f(x) \Leftrightarrow x = f^{-1}(y) \Leftrightarrow y = (f^{-1})^{-1}(x)$, we can see:

$y = f(x)$ and also, $y = (f^{-1})^{-1}(x)$.

So we get: $y = f(x) = (f^{-1})^{-1}(x) \Rightarrow f(x) = (f^{-1})^{-1}(x)$. And in short: $f = (f^{-1})^{-1}$.

So the theorem that $(f^{-1})^{-1} = f$ is saying a fact that the inverse of the inverse is the original function. The fact sounds self-evident. Nevertheless, the proof itself can still be quite confusing to many students. *Simple is not necessarily easy unless understood.*

We can hardly see things not only too far but too close, too.

So let's this time, take a closer look in a bit different angle.

To begin with, the definition for inverse functions is: $y = f(x) \Leftrightarrow x = f^{-1}(y)$.

So by the definition, we get: $x = f^{-1}(y) \Leftrightarrow y = (f^{-1})^{-1}(x)$.   How come?

We have used just the definition that $y = f(x) \Leftrightarrow x = f^{-1}(y)$.   Still foggy?

First, taking the inverse of f, we use a notation where f^{-1}, read as: f inverse.

And next, the statement that $y = f(x) \Rightarrow x = f^{-1}(y)$ is saying that taking the inverse of a function $y = f(x)$, we get: $x = f^{-1}(y)$, which is thus, the inverse of f.

And the statement that $y = f(x) \Leftarrow x = f^{-1}(y)$ is saying that taking the original function of the inverse $x = f^{-1}(y)$, we get: $y = f(x)$, which is therefore, the original function.

And also, of course, taking the inverse of the inverse, we get the original function back. That is, $(f^{-1})^{-1} = f$, which is the original.

And we can put the idea this way: $x = f^{-1}(y) \Leftrightarrow y = (f^{-1})^{-1}(x)$, where $(f^{-1})^{-1} = f$.

So we get: $x = f^{-1}(y) \Leftrightarrow y = f(x)$.

Specifically though, what can we mean by this: $y = f(x) \Rightarrow x = f^{-1}(y)$?

First, of the inverse function f^{-1}, all the inputs are all the outputs of f.

And we know we use y as the output variable in f, and y gets every output of f. So we use y as the input variable in f^{-1}.

And next, all the outputs of f^{-1} are all the inputs of the original function f.

And we know we use x as the input variable in f, and x gets every input of f.

So we use x as the output variable in f^{-1}.

(Functions are not easy, much less the inverses. It is often the case though, once understood, it's quite obvious.)

- And moving now, on to the theorem 1, we have: $f^{-1}\{f(x)\} = x$.   How come?

To begin with, we have the definition where: $y = f(x) \Leftrightarrow x = f^{-1}(y)$.

So we get: $f^{-1}(y) = f^{-1}\{f(x)\}$ if putting $f(x)$ into y in $f^{-1}(y)$, since $y = f(x)$.

And thus, we get: $x = f^{-1}(y) = f^{-1}\{f(x)\} \Rightarrow x = f^{-1}\{f(x)\}$, which is, for short: $x = f^{-1}(f)$.

- And moving next, on to the theorem 2, we have: $f\{f^{-1}(y)\} = y$.

Then again first, we have the definition: $y = f(x) \Leftrightarrow x = f^{-1}(y)$.

So we get: $f(x) = f\{f^{-1}(y)\}$ if putting $f^{-1}(y)$ into x in $f(x)$, since $x = f^{-1}(y)$.

And thus, we get: $y = f(x) = f\{f^{-1}(y)\}$. Therefore, we get: $y = f\{f^{-1}(y)\}$.

• Let's next, move on to the theorem 3, which is saying: $y = f(x) = x \Rightarrow f^{-1}(f) = f(f^{-1})$.

First, we have the definition: $y = f(x) \Leftrightarrow x = f^{-1}(y)$.

Thus, we can see that:

$x = f^{-1}(y) = f^{-1}\{f(x)\}$ if putting $f(x)$ into y in $f^{-1}(y)$ since $y = f(x)$, so we get: $x = f^{-1}\{f(x)\}$.

And also, we can see that $y = f(x) = f\{f^{-1}(y)\}$ since $x = f^{-1}(y)$, so we get: $y = f\{f^{-1}(y)\}$.

Besides, we have: $y = x$ since $y = f(x) = x$. So, we get: $f\{f^{-1}(y)\} = f^{-1}\{f(x)\}$.

And thus, we get this, too: $f^{-1}\{f(x)\} = f^{-1}\{f(y)\}$ since $y = x$.

Thus, we now have: $f^{-1}\{f(x)\} = f^{-1}\{f(y)\}$, and $f\{f^{-1}(y)\} = f^{-1}\{f(x)\}$.

So we get: $f^{-1}\{f(y)\} = f\{f^{-1}(y)\}$.

And also, since $y = x$, we get: $f^{-1}\{f(y)\} = f\{f^{-1}(y)\} \Rightarrow f^{-1}\{f(x)\} = f\{f^{-1}(y)\}$.

Therefore, we can now say that $f^{-1}(f) = f(f^{-1})$ if f is an identity function as $y = f(x) = x$.

• Now, let's next, move on to the theorem 4, which is saying that:

Assuming I_x is an identity function of x, and I_y is an identity function of y, we get:
$I_x = f^{-1} \bullet f$, $I_y = f \bullet f^{-1}$, and $(g \bullet f = I_x$ and $f \bullet g = I_y) \Leftrightarrow g = f^{-1}$.

The theorem above is in two parts, and one of the two is as below:

Assuming I_x is an identity function of x, and I_y is an identity function of y, we get:
$I_x = f^{-1} \bullet f$, and $I_y = f \bullet f^{-1}$.

• Beginning with $I_x = f^{-1} \bullet f$, we can set: $I_x = I(x) = x$, since I_x is an identity function of x.

So we get: $x = f^{-1} \bullet f$, which means therefore, the composite function $f^{-1} \bullet f$ is an identity function of x, too.

Now, we have: $y = f(x) \Leftrightarrow x = f^{-1}(y)$, and we can put it this way, too: $y = f \Leftrightarrow x = f^{-1}$.

So to begin with, we can set: $f^{-1} \bullet f = (f^{-1} \bullet f)(x) = f^{-1}(f)$. And we know: $y = f$.

Thus, we get: $f^{-1} \bullet f = f^{-1}(f) = f^{-1}(y)$, which is x since we have: $x = f^{-1}(y)$.

So we get: $f^{-1} \bullet f = f^{-1}(y) = x \Rightarrow f^{-1} \bullet f = x$, and thus, we get: $I_x = f^{-1} \bullet f$.

   • And next, moving on to $I_y = f \bullet f^{-1}$, we can set: $I_y = I(y) = y$, since I_y is an identity function of y.

So we get: $y = f \bullet f^{-1}$, which means therefore, the composite function $f \bullet f^{-1}$ is an identity function of y, too.

Now, we have: $y = f(x) \Leftrightarrow x = f^{-1}(y)$, and we can put it this way, too: $y = f \Leftrightarrow x = f^{-1}$.

So first, we can set: $f \bullet f^{-1} = (f \bullet f^{-1})(y) = f(f^{-1})$. And we have: $x = f^{-1}$.

Thus, we get: $f \bullet f^{-1} = f(f^{-1}) = f(x)$, which is y since we have: $y = f(x)$.

So we get: $f \bullet f^{-1} = f(x) = y \Rightarrow f \bullet f^{-1} = y$, and thus, we get: $I_y = f \bullet f^{-1}$.

• Now, let's next, move on to the second part in the theorem 4, and the second part is:

Assuming I_x is an identity function of x, and I_y is an identity function of y, we get: $g \bullet f = I_x$ and $f \bullet g = I_y \Leftrightarrow g = f^{-1}$.

Proving a statement that has this sign: $\Leftrightarrow$, which is read as: if and only if, we need to show that $\Rightarrow$ is true, and also, that $\Leftarrow$ is true. So we want to prove two statements below:

$g \bullet f = I_x$ and $f \bullet g = I_y \Rightarrow g = f^{-1}$.

$g = f^{-1} \Rightarrow g \bullet f = I_x$ and $f \bullet g = I_y$.

So let's now begin with: $g \bullet f = I_x$ and $f \bullet g = I_y \Rightarrow g = f^{-1}$.

First, we have: $g \bullet f = I_x$ and $f \bullet g = I_y$.

And also, we have the fact that $I_x = f^{-1} \bullet f$, and $I_y = f \bullet f^{-1}$.

So we get: $g \bullet f = I_x = f^{-1} \bullet f$, and $f \bullet g = I_y = f \bullet f^{-1}$.

Thus, we get: $g \bullet f = f^{-1} \bullet f$, and $f \bullet g = f \bullet f^{-1}$. So we can see that: $g = f^{-1}$.

And thus, we can conclude that $g \bullet f = I_x$ and $f \bullet g = I_y \Rightarrow g = f^{-1}$.

Now, we know: $I_x = I(x) = x$, and $I_y = I(y) = y$, so we have: $g \bullet f = x$, and $f \bullet g = y$.

So we may want to put the conclusion the way below:

$g \bullet f = x$, and $f \bullet g = y \Rightarrow g = f^{-1}$, which means g is the inverse of f.

And also, of course, f is the inverse of g.

    • Now, let's next, move on to the other way: $g = f^{-1} \Rightarrow g \bullet f = I_x$ and $f \bullet g = I_y$.

First, since $g = f^{-1}$, we can get: $g \bullet f = f^{-1} \bullet f$ and $f \bullet g = f \bullet f^{-1}$.

And we have this, too: $I_x = f^{-1} \bullet f$, and $I_y = f \bullet f^{-1}$.

So we get: $g \bullet f = I_x$ and $f \bullet g = I_y$.

Therefore, $g = f^{-1} \Rightarrow g \bullet f = I_x$ and $f \bullet g = I_y$.
In other words:
If $y = f(x)$ and $g = f^{-1}$, we get: $g \bullet f = x$, and $f \bullet g = y$.

30

• Now, let's move on to another theorem, and the theorem is saying that:

If $z = h(y)$ is one-to-one, we get: $(h \bullet f)^{-1} = f^{-1} \bullet h^{-1}$.

First, we know that the theorem above is saying that if $g \bullet f$ and $f \bullet g$ are identity functions, g is the inverse of f.

That is to say that: $g \bullet f = x$, and $f \bullet g = y \Rightarrow g = f^{-1}$, which means g is the inverse of f.

• What then, can we mean by $(h \bullet f)^{-1} = f^{-1} \bullet h^{-1}$?

Of course, $f^{-1} \bullet h^{-1}$ is a composite function. More importantly, it is the inverse of $h \bullet f$.

So showing that $(f^{-1} \bullet h^{-1}) \bullet (h \bullet f)$ and $(h \bullet f) \bullet (f^{-1} \bullet h^{-1})$ are identity functions, we have shown that $(h \bullet f)^{-1} = f^{-1} \bullet h^{-1}$.

And looking at $(f^{-1} \bullet h^{-1}) \bullet (h \bullet f)$ and $(h \bullet f) \bullet (f^{-1} \bullet h^{-1})$, we can think of another theorem, covered in the section, **Composite Functions**. And the theorem is as follows:

$r \bullet (q \bullet p) = (r \bullet q) \bullet p = r \bullet q \bullet p$.

That is to say that composite-function-operations are associative.
And the theorem above can be called the composite identity.

• Now, let's first show that $(f^{-1} \bullet h^{-1}) \bullet (h \bullet f)$ is an identity function.

Then, to begin with, we can set: $(f^{-1} \bullet h^{-1}) \bullet (h \bullet f) = f^{-1} \bullet (h^{-1} \bullet h) \bullet f$ by the identity above.

And also by the theorem 4 above, we can set: $h^{-1} \bullet h = I$, where I is an identity function.

Besides, we have: $k \bullet I = k$, where k is a function, of course.

Thus, we get: $(f^{-1} \bullet h^{-1}) \bullet (h \bullet f) = f^{-1} \bullet I \bullet f = f^{-1} \bullet f = I$ by the theorem 4 above.

Why not I_y or I_x but just I, though?

Precisely, we have to use I_y or I_x, and yet, we can just use I for simplicity.

We have the definition: $y = f(x) \Leftrightarrow x = f^{-1}(y)$, so f^{-1} is a function of y.

So we can set: $h^{-1} \bullet h = I_y$, where $z = I(y) = y$, which is an identity function of y.

Also, we can get the same by the theorem 1, which is $f^{-1}(f) = x$.

Note that in the function $z = I_y$, z is the output variable, and y is the input variable.

• Now next, let's show that $(h \bullet f) \bullet (f^{-1} \bullet h^{-1})$ is an identity function.

Then, to begin with, we can set: $(h \bullet f) \bullet (f^{-1} \bullet h^{-1}) = h \bullet (f \bullet f^{-1}) \bullet h^{-1}$ by the composite identity above. And also by the theorem 4 above, we can set: $f \bullet f^{-1} = I$.

Besides, we have: $k \bullet I = k$, where k is a function, of course.

Thus, we get: $(h \bullet f) \bullet (f^{-1} \bullet h^{-1}) = h \bullet I \bullet h^{-1} = h \bullet h^{-1} = I$ by the theorem 4 above.

Now, putting threads together, we have: $(f^{-1} \bullet h^{-1}) \bullet (h \bullet f) = I$, and $(h \bullet f) \bullet (f^{-1} \bullet h^{-1})$ I.

And therefore, we get: $(h \bullet f)^{-1} = f^{-1} \bullet h^{-1}$.

And working with functions, we often put them in a graph.
Putting a function in a graph, we construct the curve of it.
That is, putting a function in a graph, we put its curve in a graph.

And quite readily, we can do so if the function is an inverse.
It is the case though, only if the curve of the original function is put in a graph already.

That's because the curve of the inverse is symmetric to the original's about the line $y = x$.

And a curve is a set of points in a coordinate plane as the *x-y* plane.
So we can put the idea the way below:

> Suppose $f: X \longrightarrow Y$, $y = f(x)$, f is one-to-one, and its curve is the set of points below:
> $\{(x, f(x))| \, x \in X\}$, which equals to $\{(x, y)| \, x \in X \text{ and } y \in Y\}$.
>
> Then, the curve of f^{-1} is: $\{(f(x), x)| \, x \in X\}$, that is, $\{(y, x)| \, x \in X \text{ and } y \in Y\}$.

Now, an inverse function is a function, too, so we can name it differently if we want to.
And we can use any letters as the variables as long as their relation is maintained.

So suppose now, $y = g(x)$ is the inverse of f defined above.
Then, the input variable in g is x, and y is the output variable.

So in the inverse function g, the input variable x gets each output of the original f.

Suppose for instance, in $f(x)$, x get 1 as an input, and 2 is the output for 1.
Then, a point (1, 2) gets produced by f.

Next, moving on to $g(x)$, we can see that x gets 2 as an input, which is the output of f.
Then, $g(2) = 1$, which is the input of f, and is assigned to y, which is the output variable
in the inverse function g. So a point (2, 1) gets produced by g.

That is, the point (1, 2) in the curve of f corresponds to the point (2, 1) in the curve of g.

Fig. 0

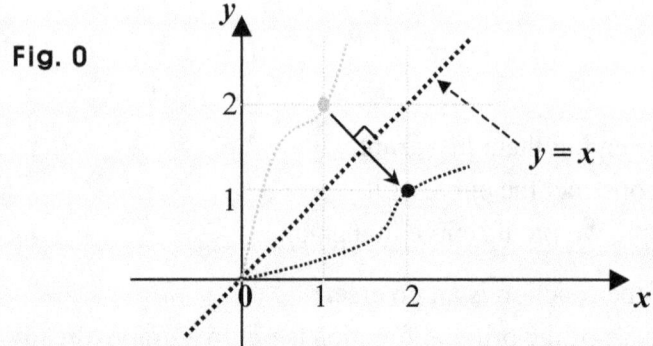

So the two points are
symmetric to each other
about the line $y = x$.

And the same is true, too, for all the points in f and g.

So if the curve of f is a set of points as follows: $\{(x, f(x)) \mid x \in X\}$, then $\{(f(x), x) \mid x \in X\}$ is the curve of g.

Therefore, what happens between f and g is as follows:

$$y = f(x) \Leftrightarrow x = f^{-1}(y) \Rightarrow y = g(x).$$

So $g = f^{-1}$, y in $y = g(x)$ gets the x-value in $f(x)$, and x in $g(x)$ gets the y-value in $y = f(x)$.

And thus, we can put the two curves in the same graph the way below:

Fig. 1

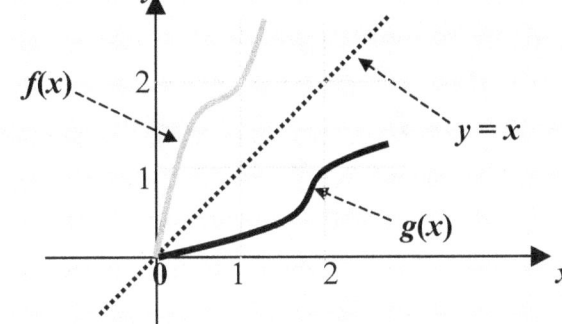

The two curves are symmetric about the line $y = x$.

So two corresponding points from the two curves make a pair, and the two points in each pair are symmetric about the line $y = x$.

The analytic approach (algebraic approach) to how all the pairs of points are symmetric about the line $y = x$ is covered in the book, **GRAPH OPERATIONS**, which covers how to manipulate curves of functions and equations.

Examples 2 in Inverse Functions

Find the inverses of the functions as follows.

0. $y = f(x) = \frac{2}{1-2x} + 1$ for $x \neq \frac{1}{2}$.

1. $y = f(x) = 2(x - 3)^3$ for x real.

2. $y = f(x) = (x - 1)(x - 2)(x - 3)$ for $x \geq 0$.

3. $y = f(x) = (x - 1)^4$ for x real.

Suggestions or Solutions
To the **Problem** in the Example **0**

Find the inverse of $y = f(x) = \frac{2}{1-2x} + 1$ for $x \neq \frac{1}{2}$.

Normally, if the expression is a fractional expression as the one above, the function is one-to-one. So the function f can be one-to-one. And thus, we may want to give it a check.

Using of two arbitrary outputs, and doing a division with the two, we can quickly check to see if it is one-to-one. If the quotient is 1, it is one-to-one.

And in fact, the curve of f is a hyperbola, so f will probably be one-to-one.
And putting the curve in a graph, we can put it the way below.

Fig. 0.0

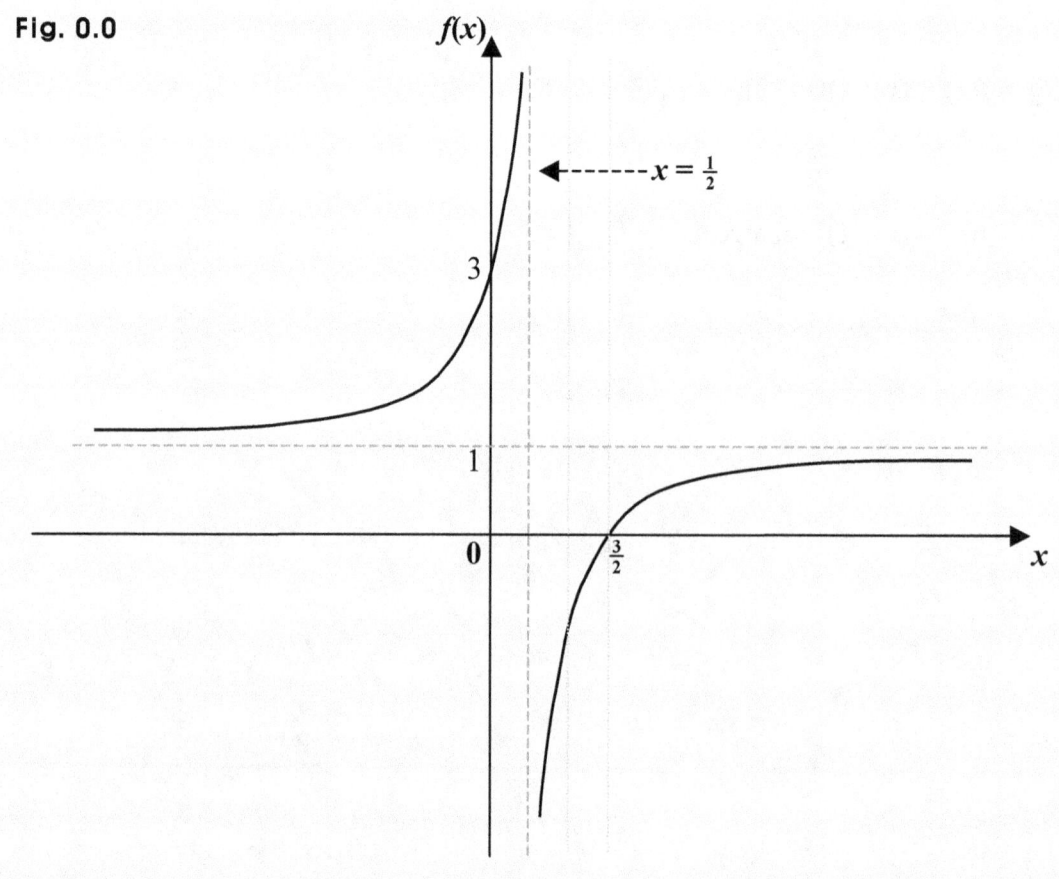

So the domain in f is a set of all real numbers less a number $\frac{1}{2}$.

That is, if R is a set of all real numbers, the domain is: $R - \{\frac{1}{2}\}$.

What about the range in f, though?

The range in f is a set of all real numbers less a number 1, so it is: $R - \{1\}$.

And we can simply put is this way, too: $y \neq 1$.

That's because y cannot be 1, because $\frac{2}{1-2x}$ cannot be 0 for any value of x, so $\frac{2}{1-2x}+1$ cannot be 1. And the domain cannot include $\frac{3}{2}$ since division by 0 is not allowed.

Since f is one-to-one, we should be able to find the inverse.

Finding the expression first, we want to solve for x the equation made of the expression of f, then swap the variables.

So solving the equation for x, we get:

$$y = \tfrac{2}{1-2x}+1 \Rightarrow y-1 = \tfrac{2}{1-2x} \Rightarrow 1-2x = \tfrac{2}{y-1} \Rightarrow 2x = 1 - \tfrac{2}{y-1} \Rightarrow x = \tfrac{1}{2} - \tfrac{1}{y-1}.$$

And next, swapping the variables, we get: $y = \tfrac{1}{2} - \tfrac{1}{x-1}$.   What then, is the next?

We want to get the domain of the inverse.

The domain is the range of f, and we know the range, which is: $y \neq 1$, that is, a set of all real numbers less 1.
So the domain in the inverse is: $x \neq 1$, which is a set of all real numbers less 1, too.

And thus, assuming g is the inverse, we get: $y = g(x) = \tfrac{1}{2} - \tfrac{1}{x-1}$ for $x \neq 1$.

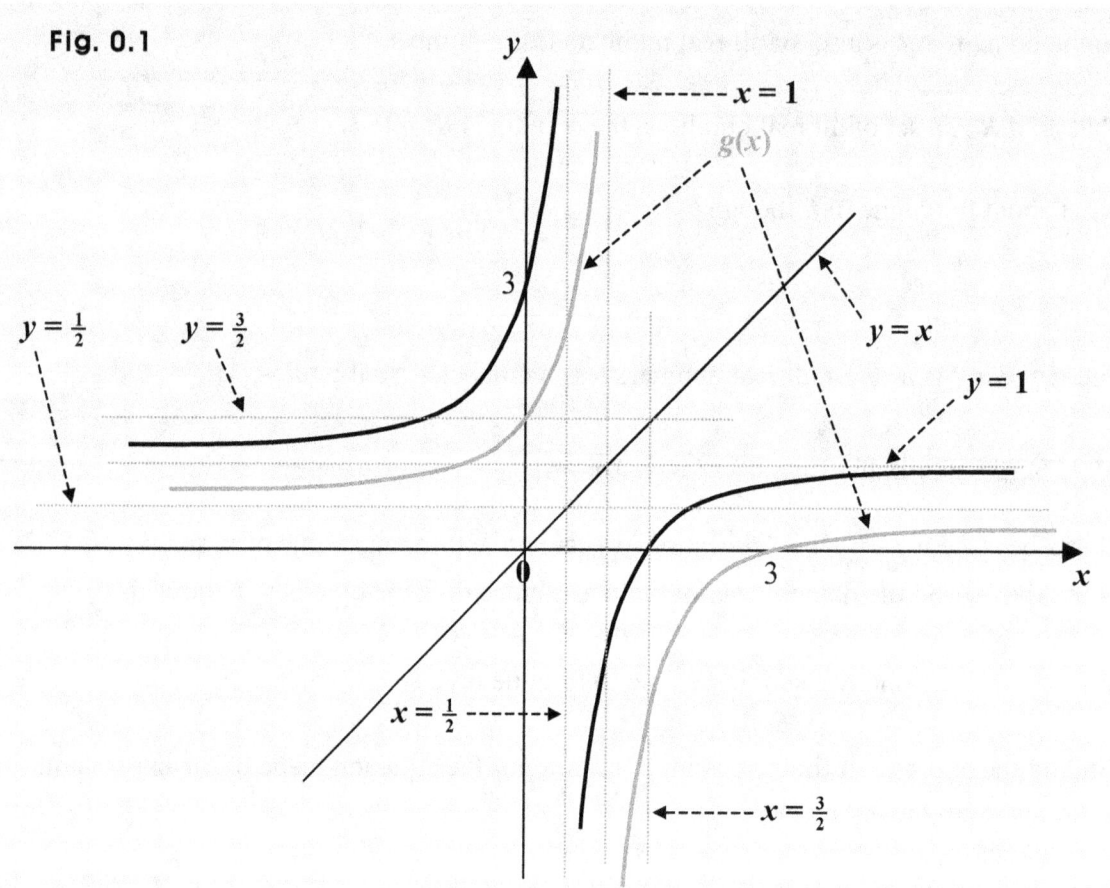

Fig. 0.1

The curve of *f* is the curve in black, and the curve of *g* is the curve in gray.

Each curve is made of two pieces, called branches in the case as above.

Such a curve is called a hyperbola.

And the two curves are symmetric about the line $y = x$.

Suggestions or Solutions
To the **Problem** in the Example **1**

Find the inverse of $y = f(x) = 2(x - 3)^3$ for x real.

Assuming $a \neq b$, a and b are constant, we get: $\frac{f(a)}{f(b)} = \frac{2(a-3)^3}{2(b-3)^3} = \frac{(a-3)^3}{(b-3)^3} = \left(\frac{a-3}{b-3}\right)^3 \neq 1.$

So the function f is one-to-one. Thus next, we can get the inverse, and get first:

$$y = 2(x - 3)^3 \Rightarrow (x - 3)^3 = \tfrac{y}{2} \Rightarrow x - 3 = \sqrt[3]{\tfrac{y}{2}} \Rightarrow x = \sqrt[3]{\tfrac{y}{2}} + 3.$$

And thus, assuming g is the inverse, we get: $y = \sqrt[3]{\tfrac{x}{2}} + 3$ for x real.

If not quite sure of the idea behind the processes above, follow the steps below:

Assuming $a \neq b$, a and b are constant, and checking to see if f is one-to-one, we get:

$\frac{f(a)}{f(b)} = \frac{2(a-3)^3}{2(b-3)^3} = \frac{(a-3)^3}{(b-3)^3} = \left(\frac{a-3}{b-3}\right)^3$, and $\frac{a-3}{b-3} \neq 1$, because $a \neq b$.

So the function f is one-to-one, and thus, we should be able to find the inverse.

Finding the expression first, we want to solve for x the equation made of the expression of f, then swap the variables. So solving the equation for x, we get:

$$y = 2(x - 3)^3 \Rightarrow (x - 3)^3 = \tfrac{y}{2} \Rightarrow x - 3 = \sqrt[3]{\tfrac{y}{2}} \Rightarrow x = \sqrt[3]{\tfrac{y}{2}} + 3.$$

And next, swapping the variables, we get: $y = \sqrt[3]{\tfrac{x}{2}} + 3 = \left(\tfrac{x}{2}\right)^{\frac{1}{3}} + 3.$

And the domain is the range of the original function f, which is a set of all real numbers.

And thus, assuming g is the inverse, we get: $y = \sqrt[3]{\tfrac{x}{2}} + 3$ for x real.

Fig. 1.0

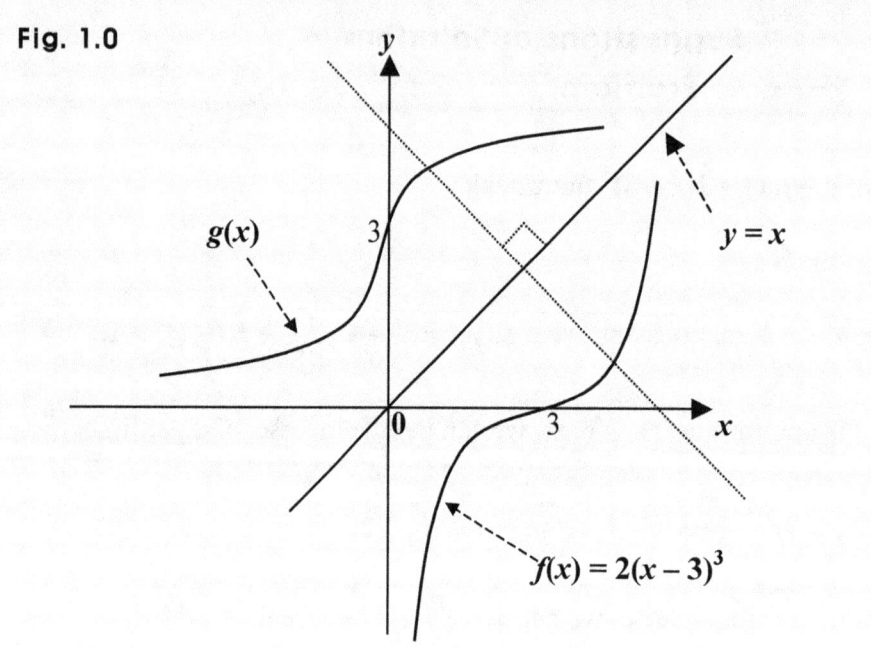

In short:

Assuming $a \neq b$, a and b are constant, we get: $\frac{f(a)}{f(b)} = \frac{2(a-3)^3}{2(b-3)^3} = \frac{(a-3)^3}{(b-3)^3} = \left(\frac{a-3}{b-3}\right)^3 \neq 1$.

So the function f is one-to-one. Thus next, we can get the inverse, and get first:

$$y = 2(x-3)^3 \Rightarrow (x-3)^3 = \frac{y}{2} \Rightarrow x - 3 = \sqrt[3]{\frac{y}{2}} \Rightarrow x = \sqrt[3]{\frac{y}{2}} + 3.$$

And thus, assuming g is the inverse, we get: $y = \sqrt[3]{\frac{x}{2}} + 3$ for x real.

Suggestions or Solutions
To the **Problem** in the Example **2**

Find the inverse of $y = f(x) = (x - 1)(x - 2)(x - 3)$ for $x \geq 0$.

The function f is not one-to-one.   How come?

The curve of f passes through the x-axis at $x = 1, 2$, and 3, which are thus, called roots. And the domain includes the roots, so f is not one-to-one.   How come?

If a and b are two different constants, and $f(a) = f(b)$, then f is not one-to-one.

Setting $a = 1$, and $b = 2$, we get: $f(1) = f(2) = 0$.   So f is not one-to-one.

Putting in a graph the curve of f, along with two horizontal lines tangent to the curve, we get:

Fig. 2.0

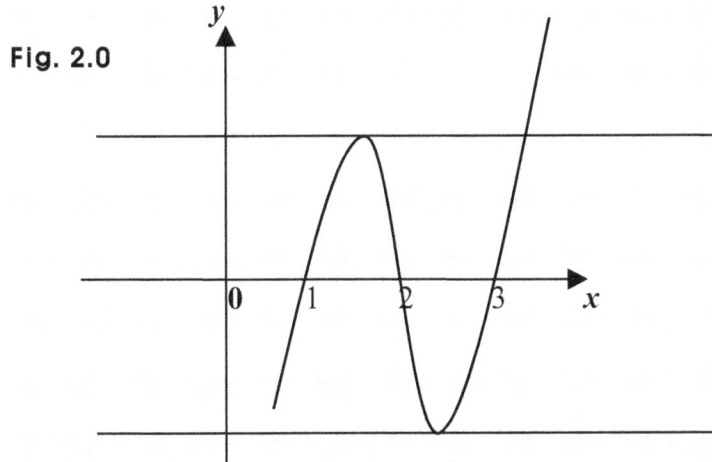

Then, finding the two points where the two lines are tangent to the curve, we can partition the domain, and thus, can try finding the inverse for each partition.

There are three partitions, but finding the two points is not quite easy.

Just doing algebra, we will probably see calculations will be quite messy.

Doing calculus though, we can readily find the two points.

However, even if we get the two points, finding the expression is still not quite easy.

That's because we have to solve an equation made of a polynomial of third degree.

How to get a specific solution algebraically to such an equation is covered in the book, **EQUATIONS**. Such a specific solution is however, in numbers, and not in an expression.

Suggestions or Solutions
To the **Problem** in the Example **3**

Find the inverse of $y = f(x) = (x-1)^4$ for x real.

Suppose a, b, s and t are constant, $a \neq b$, $s = a - 1$, and $t = b - 1$.

Then, $s \neq t$, and we get: $f(a) - f(b) = (a-1)^4 - (b-1)^4 = s^4 - t^4 = (s^2 - t^2)(s^2 + t^2)$
$= (s-t)(s+t)(s^2 + t^2)$, which can be 0.

Thus, f is not one-to-one. However, we can partition f into two functions as below:
One is: $y = f_1(x) = (x-1)^4$ for $x \geq 1$. And the other is: $y = f_2(x) = (x-1)^4$ for $x < 1$.

If $a \geq 1$, $b > 1$, and $a = 2 - b$, we get: $a \geq 1 \Rightarrow 2 - b \geq 1 \Rightarrow b \leq 1$, but we have: $b > 1$.
So consequently, we get: $f(a) - f(b) \neq 0$. And thus, $f_1(x)$ is one-to-one.

If $a < 1$, $b < 1$, and $a = 2 - b$, we get: $a < 1 \Rightarrow 2 - b < 1 \Rightarrow b > 1$, but we have: $b < 1$.
So consequently, we get: $f(a) - f(b) \neq 0$. And thus, $f_2(x)$ is one-to-one.

So to begin with: $y = (x-1)^4 \Rightarrow x - 1 = \pm\sqrt[4]{y} \Rightarrow x = \pm\sqrt[4]{y} + 1$.

Next finding the range of f_1, we get: $x \geq 1 \Rightarrow x - 1 \geq 0 \Rightarrow (x-1)^4 \geq 0$.

And next, finding the range of f_2, we get: $x < 1 \Rightarrow x - 1 < 0 \Rightarrow (x-1)^4 > 0$.

So next, checking the domains, we get:
$y \geq 0 \Rightarrow \sqrt[4]{y} \geq 0 \Rightarrow \sqrt[4]{y} + 1 \geq 1$, which is the domain of f_1.
$y > 0 \Rightarrow \sqrt[4]{y} > 0 \Rightarrow -\sqrt[4]{y} < 0 \Rightarrow -\sqrt[4]{y} + 1 < 1$, which is the domain of f_2.

And we know the range of f_1 is: $y \geq 0$, and the range of f_2 is: $y > 0$.
And thus, assuming g_1 is the inverse of f_1, we get: $y = g_1(x) = \sqrt[4]{x} + 1$ for $x \geq 0$.
And assuming g_2 is the inverse of f_2, we get: $y = g_2(x) = -\sqrt[4]{x} + 1$ for $x > 0$.

If not quite sure of the idea behind the processes above, follow the steps below:

Let's check first, to see if the function f is one-to-one.

Then, assuming a and b are two different constants, we can check to see if we get: $f(a) - f(b) \neq 0$.

To begin with, we get: $f(a) - f(b) = (a-1)^4 - (b-1)^4$, which looks a bit bulky.

So setting $s = a - 1$, and $t = b - 1$, we get first: $s \neq t$, so $s - t \neq 0$, since $a \neq b$.

And next, we get: $f(a) - f(b) = s^4 - t^4 = (s^2 - t^2)(s^2 + t^2) = (s - t)(s + t)(s^2 + t^2)$.

We know: $s - t \neq 0$, and also, $(s^2 + t^2) \neq 0$ because s and t cannot be 0 at the same time since $s \neq t$.

However, we can get: $s + t = 0$ if $s = -t$ or $-s = t$, of course.

We have: $s = a - 1$, and $t = b - 1$, so we get: $s + t = a + b - 2 = 0$ if $a = 2 - b$ or $b = 2 - a$.

For instance, if $a = 0$ and $b = 2$, we get: $s + t = a + b - 2 = 0$.

And thus, we can get: $f(a) - f(b) = 0$ for some values of a and b, that is, we cannot get: $f(a) - f(b) \neq 0$ for all values of a and b. So f is not one-to-one.

And putting the curve of f in a graph, we can get one as follows:

Fig. 3.0

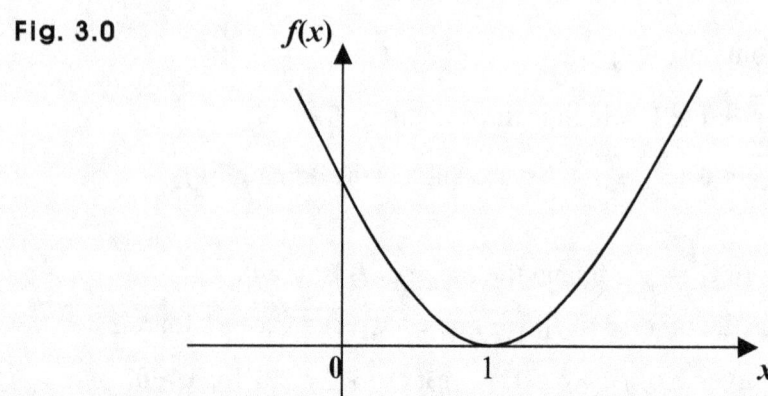

We can try partitioning though, the domain of f so that we can try getting the inverse for each partition. And partitioning, we in fact, partition the function f.
We can partition f into two functions as below:

One is: $y = f_1(x) = (x-1)^4$ for $x \geq 1$. And the other is: $y = f_2(x) = (x-1)^4$ for $x < 1$.

Then, both functions f_1 and f_2 are one-to-one. Let's see though, how it is the case.

We have already done much of the work, and the question is: Can we get: $a + b - 2 \neq 0$?
That is, we want check to see if we can get this: $a \neq 2 - b$.

So beginning with the function f_1, we have: $a \geq 1$, and $b > 1$ since the domain is: $x \geq 1$.
Why not $b \geq 1$, though?

That's because we need to have: $a \neq b$.
So we can put it this way, too: $a > 1$, and $b \geq 1$.

Now, suppose $a \geq 1$, $b > 1$, we can get: $a = 2 - b$, and $a = 1$.   What then, is b?

If $a = 1$, we get: $a = 2 - b \Rightarrow 1 = 2 - b \Rightarrow b = 1$, which is however, not allowed since we have: $a \neq b$. So there is no value for b.

And in fact, we cannot get: $a = 2 - b$ if $a \geq 1$, and $b > 1$.   How come?

Assuming $a \geq 1$, $b > 1$, and $a = 2 - b$, we get:

$a \geq 1 \Rightarrow 2 - b \geq 1 \Rightarrow b \leq 1$, which is not possible because we have: $b > 1$.

So we cannot get: $a = 2 - b$ if $a \geq 1$, and $b > 1$.

And consequently, we get: $f(a) - f(b) \neq 0$. And thus, $f_1(x)$ is one-to-one.

Putting f_1 in a graph, we can get one as follows:

Fig. 3.1

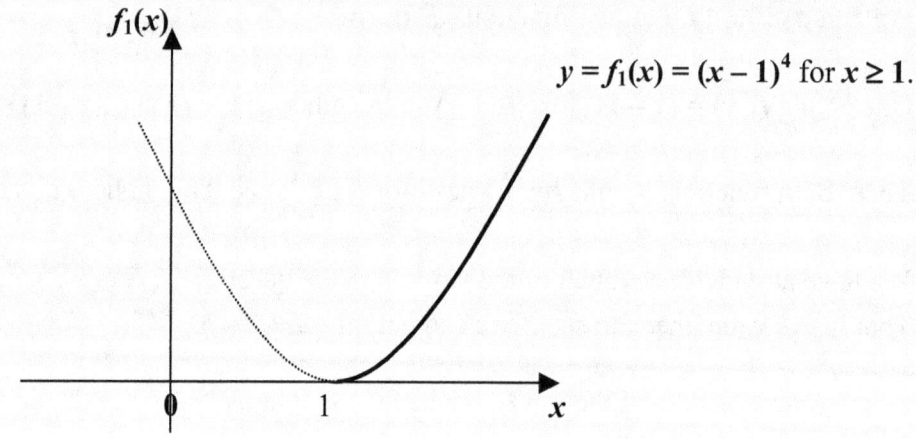

$$y = f_1(x) = (x-1)^4 \text{ for } x \geq 1.$$

And next moving on to f_2, we have: $a < 1$, and $b < 1$ since the domain is: $x < 1$.

Now, suppose $a < 1$, $b < 1$, we can get: $a = 2 - b$, and $a = 0$. What then, is b?

If $a = 0$, we get: $a = 2 - b \Rightarrow 0 = 2 - b \Rightarrow b = 2$, which is not allowed since $b < 1$.

So there is no value for b.

And in fact, we cannot get: $a = 2 - b$ if $a < 1$ and $b < 1$. How come?

Assuming $a < 1$, $b < 1$, and $a = 2 - b$, we get:

$a < 1 \Rightarrow 2 - b < 1 \Rightarrow b > 1$, which is not possible because we have: $b < 1$.

So we cannot get: $a = 2 - b$ if $a < 1$, and $b < 1$.

And consequently, we get: $f(a) - f(b) \neq 0$. And thus, $f_2(x)$ is one-to-one.

Putting f_2 in a graph, we can get one as follows:

Fig. 3.2

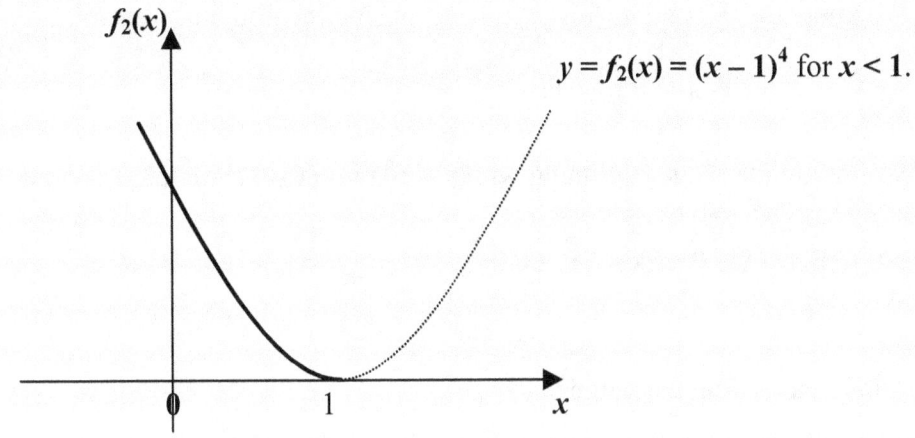

$y = f_2(x) = (x-1)^4$ for $x < 1$.

So we should be able to find the inverse of each of the two functions f_1 and f_2.

To begin with, we want to find the expression of each.

Finding the expression though, we just want to solve for x the equation made of the expression of f, then swap the variables.

That's because the two original functions f_1 and f_2 share the same expression. So solving the equation for x, we get:

$$y = (x-1)^4 \Rightarrow x - 1 = \pm\sqrt[4]{y} \Rightarrow x = \pm\sqrt[4]{y} + 1.$$

And next, swapping the variables, we get: $y = \pm\sqrt[4]{x} + 1$.

So we get two expressions, one is the expression in the inverse of f_1, and the other is the expression in the inverse of f_2.   How can we see which is which, though?

Using the range in each original function, we can make a right choice.

Before swapping variables, we had: $x = \pm\sqrt[4]{y} + 1$.

And if we begin with f_1, in the expression above, the extent of the values y can take is the range in f_1, and the extent of the values x can take is the domain in f_1.

We know the domain in f_1 is: $x \geq 1$. And using the domain, we can find the range in f_1. And the same is true for f_2, also.

So finding the range beginning with $y = f_1(x) = (x - 1)^4$ for $x \geq 1$, we get:

$$x \geq 1 \Rightarrow x - 1 \geq 0 \Rightarrow (x - 1)^4 \geq 0.$$

And next, finding the range in $y = f_2(x) = (x - 1)^4$ for $x < 1$, we get:

$$x < 1 \Rightarrow x - 1 < 0 \Rightarrow (x - 1)^4 > 0.$$

And thus, the range in f_1 is: $y \geq 0$, and the range in f_2 is: $y > 0$.

So applying the ranges to the expressions in the inverses, we get:

$y \geq 0 \Rightarrow \sqrt[4]{y} \geq 0 \Rightarrow \sqrt[4]{y} + 1 \geq 1$, which is the domain in f_1.

$y > 0 \Rightarrow \sqrt[4]{y} > 0 \Rightarrow -\sqrt[4]{y} < 0 \Rightarrow -\sqrt[4]{y} + 1 < 1$, which is the domain in f_2.

So the expression in the inverse of f_1 is: $\sqrt[4]{y} + 1$.

And the expression in the inverse of f_2 is: $-\sqrt[4]{y} + 1$.

And we know the range in f_1 is: $y \geq 0$, and the range in f_2 is: $y > 0$.

And thus, assuming g_1 is the inverse of f_1, we get: $y = g_1(x) = \sqrt[4]{x} + 1$ for $x \geq 0$.

And assuming g_2 is the inverse of f_2, we get: $y = g_2(x) = -\sqrt[4]{x} + 1$ for $x > 0$.

Putting g_1 and f_1 in one graph, we can get:

Fig. 3.3

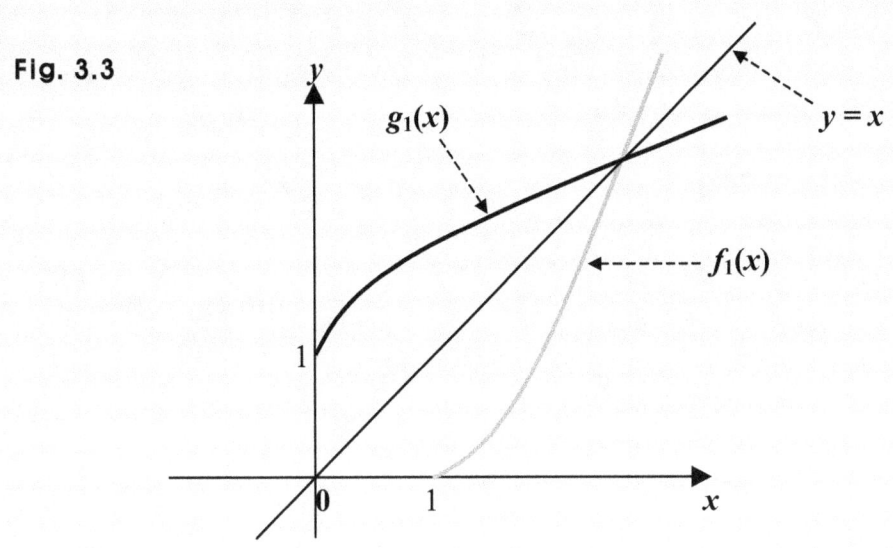

And putting g_2 and f_2 in one graph, we can get:

Fig. 3.4

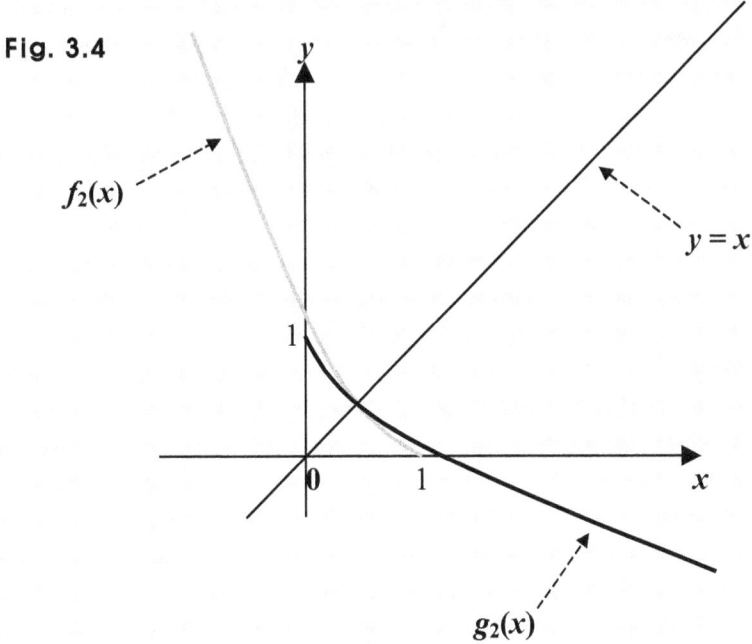

Examples 3 in Inverse Functions

0. Suppose that $f(x) = \frac{2ax+b}{cx-2}$ has its inverse as follows: $f^{-1}(x) = \frac{x+2}{2x-3}$ for $x \neq \frac{3}{2}$.

Then, find the values of a, b and c.

1. Assuming $f(x) = 2x + u$ and $g(x) = v - 3x$, find f^{-1}, g^{-1}, $g \bullet f$, $(g \bullet f)^{-1}$, $(g \bullet f) \bullet (g \bullet f)^{-1}$, and check to see if $(g \bullet f)^{-1} = f^{-1} \bullet g^{-1}$.

2. Assuming $f(x) = \sqrt{4x^2 - 1}$ for $x \geq \frac{1}{2}$, and $g(x) = x(x^2 - 2x - 1)$ for $x \geq 2$, find $f^{-1}(1)$ and $g^{-1}(-2)$.

Suggestions or Solutions
To the **Problem** in the Example **0**

Suppose that $f(x) = \frac{2ax+b}{cx-2}$ **has its inverse as follows:** $f^{-1}(x) = \frac{x+2}{2x-3}$ **for** $x \neq \frac{3}{2}$.
Then, find the values of a, b **and** c.

Assuming $y = f(x)$, we get: $y = \frac{2ax+b}{cx-2} \Rightarrow (cx-2)y = 2ax + b \Rightarrow x(cy-2a) = b + 2y$

$\Rightarrow x = \frac{2y+b}{cy-2a}$. And next, swapping the variables, we get: $y = \frac{2x+b}{cx-2a}$.

And we have: $f^{-1}(x) = \frac{x+2}{2x-3} = \frac{2x+4}{4x-6}$, which equals $\frac{2x+b}{cx-2a}$.

So comparing the numerators, we get: $b = 4$, and next, comparing term by term in the denominators, we can simply get: $c = 4$, and $a = 3$.

If not quite sure of the idea behind the processes above, follow the steps below:

This is just an example of algebra practice on inverse functions.

Getting the solution to a problem, we want to first, understand the problem, of course.

Understanding is one thing, though. Doing it is another. And doing it, we usually need to do algebra. We can actually get the solution doing algebra.

Now, we have the definition for inverse functions as follows: $y = f(x) \Leftrightarrow x = f^{-1}(y)$, where $f(x)$ is one-to-one, of course.

How come then, do we have $f^{-1}(x)$ instead of $f^{-1}(y)$? That is, why x and not y?

There is nothing wrong with the variable, and it simply means that the letter x has been chosen to be used as the input variable in the inverse of the function f. How come?

Using y as the input variable as in $f^{-1}(y)$, we can just put the expression this way: $\frac{y+2}{2y-3}$. Not quite sure?

Taking the inverse of f, we first solve for x the equation made of the expression of f, then swap the variables. And then, we usually choose a name for the inverse.

So for instance, we often set: $y = g(x)$.

Even after swapping the variables however, if we keep the name of the inverse unchanged, the name will just remain to be f^{-1} instead of g as in the instance above.

So if g is chosen to be the name of the inverse, we can just set: $y = g(x) = \frac{x+2}{2x-3}$ for $x \neq \frac{3}{2}$. What can we do about this problem, then?

We just take the inverse of f, and compare it to the one given, which is: $\frac{x+2}{2x-3}$, of course.

So assuming $y = f(x)$, and solving the equation made of the expression of f, we get:

$$y = \frac{2ax+b}{cx-2} \Rightarrow (cx-2)y = 2ax + b \Rightarrow x(cy - 2a) = b + 2y \Rightarrow x = \frac{2y+b}{cy-2a}.$$

And next, swapping the variables, we get: $y = \frac{2x+b}{cx-2a}$.

So next, we want to compare the expression above with the one given, which is: $\frac{x+2}{2x-3}$.

We now have: $\frac{2x+b}{cx-2a}$, and $\frac{x+2}{2x-3}$. Comparing the two though, we can see the coefficients of the x-terms in the numerators don't match. So we want to make both the same.

Then, we simply get: $\frac{2x+b}{cx-2a}$, and $\frac{2x+4}{4x-6}$.

Now, comparing the constant terms in the numerators, we just get: $b = 4$, and next, comparing term by term in the denominators, we simply get: $c = 4$, and $a = 3$.

And we can get the same in just another way, too:

We know that if there exists the f^{-1} of an f, we get: $(f^{-1})^{-1} = f$, which is a theorem, and is saying that the inverse of the inverse is just the original function, and the idea was covered when we covered the theorems on inverses in the section **Inverse Functions 2**.

So let's use the theorem. That is, we take the inverse of the inverse, then do the comparison. We have: $y = f^{-1}(x) = \frac{x+2}{2x-3}$. So finding the expression of the inverse of f^{-1}, we want to solve for x the equation $y = \frac{x+2}{2x-3}$, then swap the variables. And thus, solving it, we get: $y = \frac{x+2}{2x-3} \Rightarrow (2x-3)y = x + 2 \Rightarrow x(2y-1) = 3y + 2 \Rightarrow x = \frac{3y+2}{2y-1}$.

And next, swapping the variables, we get: $y = \frac{3x+2}{2x-1}$.

So the expression $\frac{3x+2}{2x-1}$ is the expression of the original function f.

We have: $f(x) = \frac{2ax+b}{cx-2}$.

So the expression in f is: $\frac{2ax+b}{cx-2}$, and thus, has to be the same as $\frac{3x+2}{2x-1}$.

And thus, we can now compare the two.

Prior to the comparison though, we may want to set: $\frac{3x+2}{2x-1} = \frac{6x+4}{4x-2}$.

Then, we can make the comparison a bit easier. We now have: $\frac{2ax+b}{cx-2}$, and $\frac{6x+4}{4x-2}$.

So comparing term by term in the numerators, we just get: $a = 3$, and $b = 4$, and next, comparing the x-terms term in the denominators, we simply get: $c = 4$.

In short:

Assuming $y = f(x)$, we get: $y = \frac{2ax+b}{cx-2} \Rightarrow (cx-2)y = 2ax + b \Rightarrow x(cy - 2a) = b + 2y$

$\Rightarrow x = \frac{2y+b}{cy-2a}$. And next, swapping the variables, we get: $y = \frac{2x+b}{cx-2a}$.

And we have: $f^{-1}(x) = \frac{x+2}{2x-3} = \frac{2x+4}{4x-6}$, which equals $\frac{2x+b}{cx-2a}$.

So comparing the numerators, we get: $b = 4$, and next, comparing term by term in the denominators, we simply get: $c = 4$, and $a = 3$.

Suggestions or Solutions

To the **Problem** in the Example **1**

Assuming $f(x) = 2x + u$ and $g(x) = v - 3x$, find f^{-1}, g^{-1}, $g \bullet f$, $(g \bullet f)^{-1}$, $(g \bullet f) \bullet (g \bullet f)^{-1}$, and check to see if $(g \bullet f)^{-1} = f^{-1} \bullet g^{-1}$.

Let's begin with the inverse function f^{-1}.

Finding the expression first, we want to solve for x the equation made of the expression of f, then swap the variables. So solving the equation for x, we get first:

$y = 2x + u \Rightarrow x = \frac{y-u}{2}$. And next, swapping the variables, we get: $y = \frac{x-u}{2}$.

And thus, $f^{-1}(x) = \frac{x-u}{2}$ for x real.

Next, moving on to the inverse of g, we get:

$y = v - 3x \Rightarrow x = \frac{v-y}{3}$. And next, swapping the variables, we get: $y = \frac{v-x}{3}$.

And thus, $g^{-1}(x) = \frac{v-x}{3}$ for x real.

Next, moving on to the composite function $g \bullet f$, we get:

$g \bullet f = g\{f(x)\} = g(f) = v - 3f = v - 3(2x + u) = v - 3u - 6x$.

Next, moving on to the inverse of the composite function $g \bullet f$, we get first:

$y = v - 3u - 6x \Rightarrow x = \frac{v-3u-y}{6}$. And next, swapping the variables, we get: $y = \frac{v-3u-x}{6}$.

And thus, $(g \bullet f)^{-1}(x) = \frac{v-3u-x}{6}$ for x real.

Let's next, move on to the composite function $(g \bullet f) \bullet (g \bullet f)^{-1}$.

Assuming $h = g \bullet f$, and $z = (g \bullet f)^{-1}$, we get:

$(g \bullet f) \bullet (g \bullet f)^{-1} = h \bullet z = h(z) = v - 3u - 6z = v - 3u - 6 \cdot \frac{v - 3u - x}{6} = v - 3u - v + 3u + x = x$.

So $(g \bullet f) \bullet (g \bullet f)^{-1}$ is an identity function.

And let's next, move on to $(g \bullet f)^{-1} = f^{-1} \bullet g^{-1}$, which is in fact, a theorem, and thus, can hold.

To begin with, $(f^{-1} \bullet g^{-1})(x) = f^{-1}(g^{-1}) = \frac{g^{-1} - u}{2}$ since $f^{-1}(x) = \frac{x - u}{2}$.

And since $g^{-1}(x) = \frac{v - x}{3}$, we get: $f^{-1}(g^{-1}) = \frac{\frac{v - x}{3} - u}{2} = \frac{\frac{v - x - 3u}{3}}{2} = \frac{v - 3u - x}{6}$, which is the same as $(g \bullet f)^{-1}(x)$.

Suggestions or Solutions
To the **Problem** in the Example **2**

Assuming $f(x) = \sqrt{4x^2 - 1}$ for $x \geq \frac{1}{2}$, and $g(x) = x(x^2 - 2x - 1)$ for $x \geq 2$, find $f^{-1}(1)$ and $g^{-1}(-2)$.

The value of $f^{-1}(1)$ is the input corresponding to the output 1 in the original function f.

We have: $y = f(x) = \sqrt{4x^2 - 1}$. So solving $1 = \sqrt{4x^2 - 1}$, we get:

$$1 = \sqrt{4x^2 - 1} \Rightarrow 4x^2 - 1 = 1 \Rightarrow x^2 = \tfrac{1}{2} \Rightarrow x = \pm\tfrac{1}{\sqrt{2}} = \pm\tfrac{\sqrt{2}}{2}.$$

And we know: $x \geq \frac{1}{2}$, so we get: $x = \frac{\sqrt{2}}{2}$, which is $f^{-1}(1)$.

The value of $g^{-1}(-2)$ is the input corresponding to the output -2 of the original function g.

We have: $y = g(x) = x(x^2 - 2x - 1)$. So solving $-2 = x(x^2 - 2x - 1)$, we get:

$$x^3 - 2x^2 - x + 2 = x^2(x - 2) - (x - 2) = (x - 2)(x^2 - 1) = (x - 2)(x - 1)(x + 1).$$

And thus, we get: $(x - 2)(x - 1)(x + 1) = 0 \Rightarrow x = -1, 1,$ or 2.

We know: $x \geq 2$. Therefore, we get: $g^{-1}(-2) = 2$.

If not quite sure of the idea behind the processes above, follow the steps below:

Let's begin with the inverse of the function f. Finding first, the expression of the inverse, we want to solve for x the equation $y = \sqrt{4x^2 - 1}$, then swap the variables.

So to begin with, $y = \sqrt{4x^2 - 1} \Rightarrow 4x^2 - 1 = y^2 \Rightarrow x = \pm\frac{1}{2}\sqrt{y^2 + 1}$.

And thus, we have two expressions to choose from. So which one?

We know the domain of f is: $x \geq \frac{1}{2}$, which means x is positive anyway, so the choice is:

$x = \frac{1}{2}\sqrt{y^2 + 1}$, since $\sqrt{y^2 + 1}$ is positive anyway. More precisely though, we can find the range of f, then see if we get the domain back applying the range to $\pm\frac{1}{2}\sqrt{y^2 + 1}$.

Finding the range, we get: $x \geq \frac{1}{2} \Rightarrow x^2 \geq \frac{1}{4} \Rightarrow 4x^2 \geq 1 \Rightarrow 4x^2 - 1 \geq 0$.

So the range is: $y \geq 0$. And thus, we get: $y \geq 0 \Rightarrow \sqrt{y^2 + 1} \geq 1 \Rightarrow \frac{1}{2}\sqrt{y^2 + 1} \geq \frac{1}{2} \Rightarrow x \geq \frac{1}{2}$.

So we get: $x = \frac{1}{2}\sqrt{y^2 + 1}$, and swapping the variables, we get: $y = \frac{1}{2}\sqrt{x^2 + 1}$.

We know that the range of f is: $y \geq 0$. So we get: $y = f^{-1}(x) = \frac{1}{2}\sqrt{x^2 + 1}$ for $x \geq 0$.

And thus, we get: $f^{-1}(1) = \frac{\sqrt{2}}{2}$. And we can get the same the way below, too:

- Directly interpreting the definition for inverse functions, we can get the solution, too.

The definition goes: $y = f(x) \Leftrightarrow x = f^{-1}(y)$, and we want to get $f^{-1}(1)$. That is, we want to get the output of the inverse function f^{-1} when $y = 1$. What then, do we mean by $y = 1$?

The y-value is an output of the original function f. And we know each output of the inverse function is each input of the original function. So the value of $f^{-1}(1)$ is the very input corresponding to the output 1 of the original function f.

We have: $y = f(x) = \sqrt{4x^2 - 1}$. So finding the x-value in the case where $1 = \sqrt{4x^2 - 1}$, we get the solution to this problem. And thus, solving $1 = \sqrt{4x^2 - 1}$, we get:

$1 = \sqrt{4x^2 - 1} \Rightarrow 4x^2 - 1 = 1 \Rightarrow x^2 = \frac{1}{2} \Rightarrow x = \pm\frac{1}{\sqrt{2}} = \pm\frac{\sqrt{2}}{2}$.

And we know: $x \geq \frac{1}{2}$, so we get: $x = \frac{\sqrt{2}}{2}$, which is $f^{-1}(1)$.

• Now, let's move on to the inverse of the other function **g**.

We have: $g(x) = x(x^2 - 2x - 1)$ for $x \geq 2$. So finding first, the expression of the inverse, we solve for x in the equation $y = x(x^2 - 2x - 1)$, then swap the variables.
How can we however, solve the equation?

It's an equation of degree three in x, so we can hardly get the solution.

However, as in the case of $f^{-1}(1)$, we don't really have to find the expression of the inverse of **g** if we just want to get $g^{-1}(-2)$.

Let's see first though, if **g** is one-to-one.

Factorizing then, $x^2 - 2x - 1$, we get: $(x - 1 - \sqrt{2})(x - 1 + \sqrt{2})$.

So we get: $g(x) = x(x - 1 - \sqrt{2})(x - 1 + \sqrt{2})$. And thus, $g = 0$ at $x = 0, 1 - \sqrt{2}$, or $1 + \sqrt{2}$.

Fig. 2.0 y $y = g(x) = x(x^2 - 2x - 1) = x(x - 1 - \sqrt{2})(x - 1 + \sqrt{2})$ for $x \geq 2$.

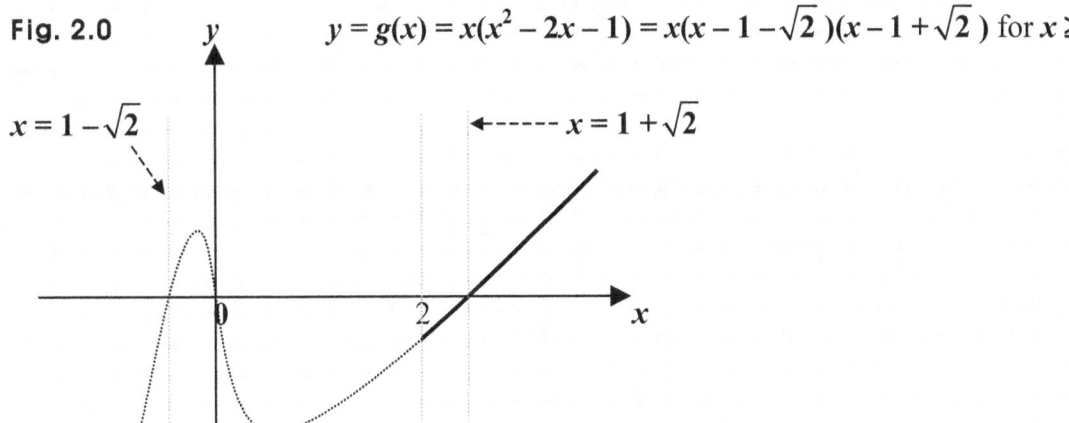

We have: $g(x) = x(x^2 - 2x - 1) = x(x - 1 - \sqrt{2})(x - 1 + \sqrt{2})$ for $x \geq 2$.

So we can actually see that **g** has its inverse since it is one-to-one.

Let's get the solution now, which is the value of $g^{-1}(-2)$.

To begin with, the definition for inverse functions is: $y = g(x) \Leftrightarrow x = g^{-1}(y)$.

So we can see that -2 in g^{-1}(-2) is an output of the original function g.

And thus, g^{-1}(-2) is the input for which the expression of g produces the output, -2.

That is, when $x = g^{-1}$(-2), $g(x) = $ -2. So in short, $g(g^{-1}$(-2)$) = $ -2.

And thus, solving for x the equation -2 $= x(x^2 - 2x - 1)$, we get the solution to this problem. So solving it, we get first:

$$-2 = x(x^2 - 2x - 1) \Rightarrow -2 = x^3 - 2x^2 - x \Rightarrow x^3 - 2x^2 - x + 2 = 0.$$

Luckily in this case, we can solve the equation factorizing the polynomial in it.

So factorizing it, we get:
$$x^3 - 2x^2 - x + 2 = x^2(x - 2) - (x - 2) = (x - 2)(x^2 - 1) = (x - 2)(x - 1)(x + 1).$$

And thus, we get: $(x - 2)(x - 1)(x + 1) = 0 \Rightarrow x = $ -1, 1, or 2.

We know: $x \geq 2$. Therefore, we get: g^{-1}(-2) $= 2$.

In short:

The value of f^{-1}(1) is the input corresponding to the output 1 of the original function f.

We have: $y = f(x) = \sqrt{4x^2 - 1}$. So solving $1 = \sqrt{4x^2 - 1}$, we get:

$$1 = \sqrt{4x^2 - 1} \Rightarrow 4x^2 - 1 = 1 \Rightarrow x^2 = \tfrac{1}{2} \Rightarrow x = \pm\tfrac{1}{\sqrt{2}} = \pm\tfrac{\sqrt{2}}{2}.$$

And we know: $x \geq \tfrac{1}{2}$, so we get: $x = \tfrac{\sqrt{2}}{2}$, which is f^{-1}(1).

The value of g^{-1}(-2) is the input corresponding to the output -2 of the original function g.

We have: $y = g(x) = x(x^2 - 2x - 1)$. So solving -2 $= x(x^2 - 2x - 1)$, we get:

$$x^3 - 2x^2 - x + 2 = x^2(x - 2) - (x - 2) = (x - 2)(x^2 - 1) = (x - 2)(x - 1)(x + 1).$$

And thus, we get: $(x - 2)(x - 1)(x + 1) = 0 \Rightarrow x = $ -1, 1, or 2.

We know: $x \geq 2$. Therefore, we get: g^{-1}(-2) $= 2$.

Examples 4 in Inverse Functions

0. Assuming $f(x) = x^2 - 3x + 2$ for $x \geq 2$, find $g(x)$ so that we get: $g\{f(x)\} = x$.

1. Assuming $(g \bullet f)(x) = 2x + 1$, $f(x) = 3x + 4$, $(u \bullet v)(x) = 2x + 1$, and $u(x) = 3x + 4$, find $g(x)$ and $v(x)$.

Suggestions or Solutions
To the **Problem** in the Example **0**

Assuming $f(x) = x^2 - 3x + 2$ for $x \geq 2$, find $g(x)$ so that we get: $g\{f(x)\} = x$.

Assuming I is an identity function, we can set: $g\{f(x)\} = (g \bullet f)(x) = I(x) = x$.

So if f is invertible, and thus, is one-to-one, we get: $g = f^{-1}$.

Assuming a and b are constant, and $a \neq b$, we can get:

$f(a) - f(b) = a^2 - 3a + 2 - (b^2 - 3b + 2) = a^2 - b^2 - 3(a-b) = (a-b)(a+b) - 3(a-b)$

$= (a-b)(a+b-3)$.   So if $a+b-3$ cannot be 0, we get: $f(a) \neq f(b)$.

Suppose now, $a \geq 2$, and $b > 2$, since $a \neq b$.   Suppose also, $a + b - 3 = 0$.

Then, we get: $a = 3 - b$. Thus, we get: $a \geq 2 \Rightarrow 3 - b \geq 2 \Rightarrow b \leq 1$, but we have: $b > 2$.

So $a + b - 3 \neq 0$, and thus, f is one-to-one. So finding f^{-1}, we can begin this way:

$y = x^2 - 3x + 2 = x^2 - 3x + (\frac{3}{2})^2 - (\frac{3}{2})^2 + 2 = (x - \frac{3}{2})^2 - (\frac{3}{2})^2 + 2 = (x - \frac{3}{2})^2 - \frac{1}{4}$.

So we get: $y + \frac{1}{4} = (x - \frac{3}{2})^2 \Rightarrow x - \frac{3}{2} = \pm\sqrt{y + \frac{1}{4}} \Rightarrow x = \pm\sqrt{y + \frac{1}{4}} + \frac{3}{2}$.

We know the domain in f is: $x \geq 2$, so finding the range in f, we get:

$x \geq 2 \Rightarrow x - \frac{3}{2} \geq \frac{1}{2} \Rightarrow (x - \frac{3}{2})^2 \geq \frac{1}{4} \Rightarrow (x - \frac{3}{2})^2 - \frac{1}{4} \geq 0$.   So the range is: $y \geq 0$.

And thus, finding the domain back, we get:

$y \geq 0 \Rightarrow y + \frac{1}{4} \geq \frac{1}{4} \Rightarrow \sqrt{y + \frac{1}{4}} \geq \frac{1}{2} \Rightarrow \sqrt{y + \frac{1}{4}} + \frac{3}{2} \geq 2 \Rightarrow x \geq 2$, which is the domain.

So we get: $x = \sqrt{y + \frac{1}{4}} + \frac{3}{2}$. And next, swapping the variables, we get: $y = \sqrt{x + \frac{1}{4}} + \frac{3}{2}$.

And thus, the inverse is: $y = f^{-1}(x) = \sqrt{x + \frac{1}{4}} + \frac{3}{2}$ for $x \geq 0$.

We know $g = f^{-1}$. So we get: $y = g(x) = \sqrt{x + \frac{1}{4}} + \frac{3}{2}$ for $x \geq 0$.

If not quite sure of the idea behind the processes above, follow the steps below:

What do we mean by $g\{f(x)\} = x$?

It is a composite function where the function f blends into the other function g.

So we can set: $g\{f(x)\} = (g \bullet f)(x) = x$, which is thus, an identity function as $y = I(x) = x$.

And thus, assuming I is an identity function, we can just set it this way, too: $g \bullet f = I$.

And we have a theorem where $f^{-1} \bullet f = I$ if f is invertible.

So if f is invertible, we can see that: $g = f^{-1}$.   What then, do we mean by f is invertible?

If f is one-to-one, f is invertible, and has its inverse.

So let's check to see if f is one-to-one.

Assuming a and b are constant, and $a \neq b$, we can check to see if $f(a) - f(b) \neq 0$, that is, $f(a) \neq f(b)$. If it is the case, f is one-to-one.

And doing the check, we need to set: $a \geq 2$, and $b > 2$, or equivalently, $a > 2$, and $b \geq 2$.

That's because the domain is: $x \geq 2$, and $a \neq b$. And thus, doing the check, we get:

$f(a) - f(b) = a^2 - 3a + 2 - (b^2 - 3b + 2) = a^2 - b^2 - 3(a - b) = (a - b)(a + b) - 3(a - b)$

$= (a - b)(a + b - 3)$.

And we know: $a - b \neq 0$ since $a \neq b$.
So if $a + b - 3$ cannot be 0, either, we get: $f(a) \neq f(b)$.

Suppose $a + b - 3 = 0$. Then, we get: $a = 3 - b$. We have: $a \geq 2$, and $b > 2$, too.

So we get: $a \geq 2 \Rightarrow 3 - b \geq 2 \Rightarrow b \leq 1$, which however, contradicts the fact that $b > 2$.

So we get: $a + b - 3 \neq 0$ if $a \geq 2$, and $b > 2$.

And the same is true, too, for the case where $a > 2$, and $b \geq 2$. That's because in this case, the case where $a \geq 2$, and $b > 2$ is equivalent to the case where $a > 2$, and $b \geq 2$.

So f is one-to-one, and thus, is invertible. And finding first, the expression of the inverse, we solve for x, the equation made of the expression of f, then swap the variables.

Without using the quadratic formula, we can begin this way:

$$y = x^2 - 3x + 2 = x^2 - 3x + \left(\tfrac{3}{2}\right)^2 - \left(\tfrac{3}{2}\right)^2 + 2 = \left(x - \tfrac{3}{2}\right)^2 - \left(\tfrac{3}{2}\right)^2 + 2 = \left(x - \tfrac{3}{2}\right)^2 - \tfrac{1}{4}.$$

So we get: $y + \tfrac{1}{4} = \left(x - \tfrac{3}{2}\right)^2 \Rightarrow x - \tfrac{3}{2} = \pm\sqrt{y + \tfrac{1}{4}} \Rightarrow x = \pm\sqrt{y + \tfrac{1}{4}} + \tfrac{3}{2}$, which means two expressions.

Thus, we need to choose one from $x = \sqrt{y + \tfrac{1}{4}} + \tfrac{3}{2}$, and $x = -\sqrt{y + \tfrac{1}{4}} + \tfrac{3}{2}$.

We know the domain of f is: $x \geq 2$, so finding the range of f, we get:

$x \geq 2 \Rightarrow x - \tfrac{3}{2} \geq \tfrac{1}{2} \Rightarrow \left(x - \tfrac{3}{2}\right)^2 \geq \tfrac{1}{4} \Rightarrow \left(x - \tfrac{3}{2}\right)^2 - \tfrac{1}{4} \geq 0.$ So the range is: $y \geq 0$.

Thus, finding the domain back, which is: $x \geq 2$, we get:

$y \geq 0 \Rightarrow y + \tfrac{1}{4} \geq \tfrac{1}{4} \Rightarrow \sqrt{y + \tfrac{1}{4}} \geq \tfrac{1}{2} \Rightarrow -\sqrt{y + \tfrac{1}{4}} \leq -\tfrac{1}{2} \Rightarrow -\sqrt{y + \tfrac{1}{4}} + \tfrac{3}{2} \leq 1 \Rightarrow x \leq 1$, which is not the domain.   So next, using the other expression, we can get:

$y \geq 0 \Rightarrow y + \tfrac{1}{4} \geq \tfrac{1}{4} \Rightarrow \sqrt{y + \tfrac{1}{4}} \geq \tfrac{1}{2} \Rightarrow \sqrt{y + \tfrac{1}{4}} + \tfrac{3}{2} \geq 2 \Rightarrow x \geq 2$, which is the domain.

So we get: $x = \sqrt{y + \tfrac{1}{4}} + \tfrac{3}{2}$. And next, swapping the variables, we get: $y = \sqrt{x + \tfrac{1}{4}} + \tfrac{3}{2}$.

And thus, the inverse is: $y = f^{-1}(x) = \sqrt{x + \tfrac{1}{4}} + \tfrac{3}{2}$ for $x \geq 0$.

And we know $g = f^{-1}$. So we get: $y = g(x) = \sqrt{x + \frac{1}{4}} + \frac{3}{2}$ for $x \geq 0$.

Fig. 3.0

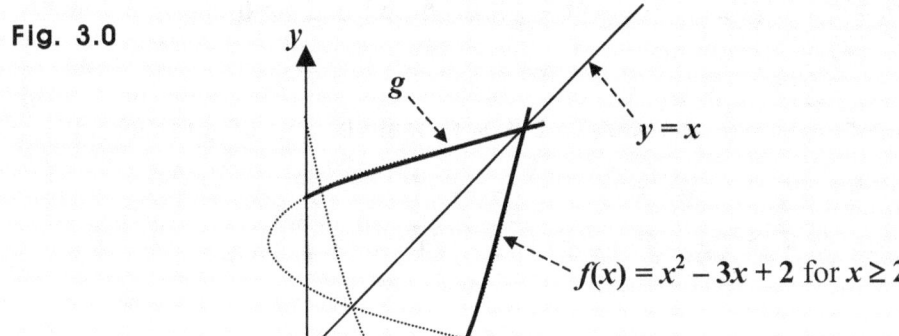

• And we can get the same, too, using the idea on composite functions.

Assuming $h = g \bullet f$, we can set: $h(x) = (g \bullet f)(x) = x$.

In other words, $h(x) = g(f(x)) = x$.

That is, putting the expression of f into the input variable in the expression of g, we get: x, which is the expression of the composite function h.

So suppose next, $t = g(s)$, where s is just the input variable, and t is merely the output variable. Then, putting the expression of f into the input variable s in the expression of g, we get: x.

That is, putting $x^2 - 3x + 2$ into the input variable s in the expression of g, we get: x.

So we can set: $s = x^2 - 3x + 2$. And thus, finding the expression of g, we get first:

$$s = x^2 - 3x + 2 = x^2 - 3x + \left(\tfrac{3}{2}\right)^2 - \left(\tfrac{3}{2}\right)^2 + 2 = \left(x - \tfrac{3}{2}\right)^2 - \left(\tfrac{3}{2}\right)^2 + 2 = \left(x - \tfrac{3}{2}\right)^2 - \tfrac{1}{4}.$$

So next, we get: $s + \frac{1}{4} = \left(x - \frac{3}{2}\right)^2 \Rightarrow x - \frac{3}{2} = \pm\sqrt{s + \frac{1}{4}} \Rightarrow x = \pm\sqrt{s + \frac{1}{4}} + \frac{3}{2}$, which means two expressions. We need thus, to choose either of the two.   How though?

We know: $x \geq 2$, and we have: $s = x^2 - 3x + 2 = (x - \frac{3}{2})^2 - \frac{1}{4}$. So we get first:

$x \geq 2 \Rightarrow x - \frac{3}{2} \geq \frac{1}{2} \Rightarrow (x - \frac{3}{2})^2 \geq \frac{1}{4} \Rightarrow (x - \frac{3}{2})^2 - \frac{1}{4} \geq 0$. So we get: $s \geq 0$.

And we have: $x = \pm\sqrt{s + \frac{1}{4}} + \frac{3}{2}$. So using: $s \geq 0$, we should be able to get this back: $x \geq 2$.

$s \geq 0 \Rightarrow s + \frac{1}{4} \geq \frac{1}{4} \Rightarrow \sqrt{s + \frac{1}{4}} \geq \frac{1}{2} \Rightarrow -\sqrt{s + \frac{1}{4}} \leq -\frac{1}{2} \Rightarrow -\sqrt{s + \frac{1}{4}} + \frac{3}{2} \leq 1 \Rightarrow x \leq 1$.

$s \geq 0 \Rightarrow s + \frac{1}{4} \geq \frac{1}{4} \Rightarrow \sqrt{s + \frac{1}{4}} \geq \frac{1}{2} \Rightarrow \sqrt{s + \frac{1}{4}} + \frac{3}{2} \geq 2 \Rightarrow x \geq 2$.

So we get: $x = \sqrt{s + \frac{1}{4}} + \frac{3}{2}$, which is the expression of g. And we have set: $t = g(s)$.

So we get: $t = g(s) = \sqrt{s + \frac{1}{4}} + \frac{3}{2}$.

We know s and t are just the input variable and the output variable in the function g.

That is, g is simply in the s-t system, which is just another name of a perpendicular coordinate system as the x-y system. In a perpendicular coordinate system where we use two variables, two coordinate axes are perpendicular to each other.

So putting g in the x-y system, we just use x and y in place of s and t, and thus, we can just set: $y = g(x) = \sqrt{x + \frac{1}{4}} + \frac{3}{2}$ for $x \geq 0$.

Suggestions or Solutions
To the **Problem** in the Example **1**

Assuming $(g \bullet f)(x) = 2x + 1$, $f(x) = 3x + 4$, $(u \bullet v)(x) = 2x + 1$, and $u(x) = 3x + 4$, find $g(x)$ and $v(x)$.

Finding f^{-1} first, we get first: $y = 3x + 4 \Rightarrow x = \frac{y-4}{3}$.

So we get: $y = f^{-1}(x) = \frac{x-4}{3}$ for x real.

Thus next, assuming $h = g \bullet f$, we get: $h(x) = 2x + 1$, and also, $h \bullet f^{-1} = g \bullet f \bullet f^{-1} = g$.

So we get: $h \bullet f^{-1} = h(f^{-1}) = 2f^{-1} + 1 = 2 \cdot \frac{x-4}{3} + 1 = \frac{2x-8+3}{3} = \frac{2x-5}{3}$.

Therefore, we get: $y = g(x) = \frac{2x-5}{3}$ for x real.

We have: $(g \bullet f)(x) = g(f(x)) = 2x + 1$. So assuming $t = g(s)$, we can set: $s = 3x + 4$.
And we know that x in $2x + 1$ is the same as the x in $3x + 4$.

And thus, finding the expression of the function g, we get:

$s = 3x + 4 \Rightarrow x = \frac{s-4}{3} \Rightarrow 2x + 1 = 2 \cdot \frac{s-4}{3} + 1 = \frac{2s-8+3}{3} = \frac{2s-5}{3}$.

So we get: $t = g(s) = \frac{2s-5}{3}$.

The two functions u and f are the same. So we have: $y = u^{-1}(x) = \frac{x-4}{3}$ for x real.

Thus, setting: $h = u \bullet v$, we get: $h(x) = 2x + 1$, and also, $u^{-1} \bullet h = u^{-1} \bullet u \bullet v = v$.

That is, $u^{-1} \bullet h = v$. We have: $u^{-1}(x) = \frac{x-4}{3}$, and $h(x) = 2x + 1$.

So we get: $u^{-1} \bullet h = u^{-1}(h) = \frac{h-4}{3} = \frac{2x+1-4}{3} = \frac{2x-3}{3}$. And thus, $y = v(x) = \frac{2x-3}{3}$ for x real.

We have: $u(x) = 3x + 1$, and $(u \bullet v)(x) = 2x + 1$.

So we can just get: $(u \bullet v)(x) = u(v(x)) = 3 \cdot v(x) + 4 = 2x + 1$.

And thus, we get: $3 \cdot v(x) = 2x + 1 - 4 = 2x - 3 \Rightarrow v(x) = \frac{2x-3}{3}$.

If not quite sure of the idea behind the processes above, follow the steps below:

Let's begin with the function **g**.

We are given two functions, one is a composite function $(g \bullet f)(x) = 2x + 1$, and the other is a function $f(x) = 3x + 4$.

In this case, we can find the function **g** using either of two ideas. One is the idea where composite function gets made, and the other is using theorems on inverse functions.

Let's begin with the theorems first. We can use two theorem as follows:

One is that $f \bullet f^{-1}$ and $f^{-1} \bullet f$ are identity functions as $y = I(x) = x$.

And the other is that $p \bullet q \bullet r = (p \bullet q) \bullet r = p \bullet (q \bullet r)$, where **p**, **q**, and **r** are functions, of course.

So putting the two theorems together, we can see that:

$$g \bullet f \bullet f^{-1} = g \bullet I = g, \quad g \bullet f^{-1} \bullet f = g \bullet I = g, \quad f \bullet f^{-1} \bullet g = I \bullet g = g, \quad \text{and } f^{-1} \bullet f \bullet g = I \bullet g = g.$$

And thus, of the four equalities above, we use this: $g \bullet f \bullet f^{-1} = g$.

So we want to find f^{-1}.

Finding the expression first, we solve for **x** the equation $y = 3x + 4$, then swap the variables. So solving the equation, we get:

$y = 3x + 4 \Rightarrow x = \frac{y-4}{3}$. And next, swapping the variables, we get: $y = \frac{x-4}{3}$.

So we get: $y = f^{-1}(x) = \frac{x-4}{3}$ for **x** real.

Thus next, assuming $h = g \bullet f$, we can set: $h(x) = (g \bullet f)(x) = 2x + 1$.

So we get: $h(x) = 2x + 1$. And also, we can get: $h \bullet f^{-1} = g \bullet f \bullet f^{-1}$, which is **g**, of course.

That is, $h \bullet f^{-1} = g$.

Now, we get: $h \bullet f^{-1} = h(f^{-1}) = 2f^{-1} + 1 = 2 \cdot \frac{x-4}{3} + 1 = \frac{2x-8+3}{3} = \frac{2x-5}{3}$.

Therefore, we get: $y = g(x) = \frac{2x-5}{3}$ for x real.

- And let's next, use the idea on composite functions.

Suppose again, $h = g \bullet f$.   Then, we can set again: $h(x) = (g \bullet f)(x) = 2x + 1$.

In other words, $h(x) = g(f(x)) = 2x + 1$.

That is, putting the expression of f into the input variable in the expression of g, we get: $2x + 1$, which is the expression of the composite function h.

So suppose next, $t = g(s)$, where s is just the input variable, and t is merely the output variable.

Then, putting the expression of f into the input variable s in the expression of g, we get: $2x + 1$.

That is, putting $3x + 4$ into the input variable s in the expression of g, we get: $2x + 1$.

Then, we can set: $s = 3x + 4$.   Now, from where is x in $2x + 1$?

The x in $2x + 1$ is from the expression of f, which is $3x + 4$.

That is, x in $2x + 1$ is the same as the x in $3x + 4$. And we have: $s = 3x + 4$.

So putting x in terms of s, we get an expression in terms of s, and then, we can put the expression into x in $2x + 1$.

That is, solving $s = 3x + 4$ for x, we get the solution which is an expression in terms of s, and then, we can put the solution into x in $2x + 1$, because x in $2x + 1$ is the same as the x in $3x + 4$. What then, do we get?

We get an expression in terms of *s*, which is the expression of the function *g*.

So putting *s* = **3x + 4** back into *s* in the expression of *g*, we get an expression in terms of *x*, and the expression is: **2x + 1**, which is the very expression of the composite function *h*.

And thus, finding the expression of *g*, we get:

$$s = 3x + 4 \Rightarrow x = \frac{s-4}{3} \Rightarrow 2x + 1 = 2 \cdot \frac{s-4}{3} + 1 = \frac{2s-8+3}{3} = \frac{2s-5}{3}.$$

So we get: $t = g(s) = \frac{2s-5}{3}$.

We know *s* and *t* are just the input variable and the output variable in the function *g*.

That is, *g* is simply in the *s-t* system, which is just another name of a perpendicular coordinate system as the *x-y* system.

So putting *g* in the *x-y* system, we just use *x* and *y* in place of *s* and *t*, and thus, we can just set: $y = g(x) = \frac{2x-5}{3}$.

• And let's next, move on to the other function *v(x)*.

As in the case above, using the two theorems, we can find the function *v(x)*.

We have: $(u \bullet v)(x) = 2x + 1$, and $u(x) = 3x + 4$.

So using the two theorems above, we can get: $u^{-1} \bullet u \bullet v = v$.

And thus, we want to find first, u^{-1}. However, we have already got the inverse because the two functions *u* and *f* are the same. So we have: $y = u^{-1}(x) = \frac{x-4}{3}$ for *x* real.

Thus, setting again: $h = u \bullet v$, we can set: $h(x) = (u \bullet v)(x) = 2x + 1$.

So we get again: $h(x) = 2x + 1$. And we can get: $u^{-1} \bullet h = u^{-1} \bullet u \bullet v$, which is *v*, of course.

That is, $u^{-1} \bullet h = v$.   We have: $u^{-1}(x) = \frac{x-4}{3}$, and $h(x) = 2x + 1$.

So we get: $u^{-1} \bullet h = u^{-1}(h) = \frac{h-4}{3} = \frac{2x+1-4}{3} = \frac{2x-3}{3}$. And thus, $y = v(x) = \frac{2x-3}{3}$ for x real.

• And let's next, find the function v using the idea on composite functions.

Suppose again, $h = u \bullet v$. Then, we can set again: $h(x) = (u \bullet v)(x) = 2x + 1$.

In other words, $h(x) = u(v(x)) = 2x + 1$.

That is, putting the expression of v into the input variable in the expression of u, we get: $2x + 1$, which is the expression of the composite function h.

In this case though, unlike the case of g above, we have the expression of the function u.

We have: $u(x) = 3x + 1$.

So we can just get: $(u \bullet v)(x) = u(v(x)) = 3 \cdot v(x) + 4 = 2x + 1$.

And thus, we get: $3 \cdot v(x) = 2x + 1 - 4 = 2x - 3 \Rightarrow v(x) = \frac{2x-3}{3}$.

₆.Exponential & Log Functions

Normally, indicating a value, we use a number as 5, 12, 1.4, 1/3, etc. We cannot indicate however, some values using such numbers. Indicating for instance, a diagonal in some square, we cannot use just a number. Indicating it, we have to put numbers in a special form, or need to use a special sign.   What sign then, is it?

It is called a radical sign as a square root sign, called a second radical sign or a second root sign, too. So for instance, if a square is 1 by 1, its diagonal is $\sqrt{2}$.

Without using such a sign, too, though, we can still indicate such a value using just numbers only. That is, we use a special expression to indicate a value using numbers. We put together numbers in a special form or structure.   So what expression is it?

It is called a power.
Putting a value in a power, we express the value in terms of a base and an exponent.
For instance, we can put the diagonal above this way, too: $2^{0.5}$.

And of course, we can put it this way, also: $2^{\frac{1}{2}}$. So we can set: $\sqrt{2} = 2^{\frac{1}{2}} = 2^{0.5}$.

Using, for another instance, the power expression, we can put 9 in 3^2, where the base is 3, and the exponent is 2. And of course, we can put it this way, too: 9^1, where the base is 9, and the exponent is 1.

And in fact, every number can be said to have a base and an exponent. For instance, we can say that a number 7 has 7 as the base, and has 1 as the exponent, since $7 = 7^1$, which is a power, a power of 7.

So technically, a number can be said to be in two parts, one is a base, and the other is an exponent. And thus, every number can be taken as a power.

Of course, it is probably the case you've already known all the fact above. It's just a warming-up.

In this section in fact, it is assumed that you have some knowledge on expressions with powers, that is, exponential expressions. And also, you are expected to be familiar with irrational numbers, radicals, and logarithms, too. For instance, you should be able to understand such numbers and expressions as follows:

$$2^{-3} = \frac{1}{2^3}, \sqrt[3]{x} \text{ or } \sqrt{x+2}, \text{ and } 2^3 2^5 = 2^{3+5} = (2^2)^4 = 2^{10-2} = \frac{2^{10}}{2^2} = 2^{\frac{16}{2}} = (2^{16})^{0.5} = (2^{1.6})^5 = \text{ etc.}$$

$$\log_2 8 = 3, \log 2 + \log 5 = \log 10 = 1, \log 1 = \log_2 1 = \log_{0.3} 1 = \log_7 1 = \log_b 1 = 0, \text{ etc.}$$

For details on exponents, powers, logarithms, radicals, and how they work, refer to **ALGEBRA EXAMPLES POWERS AND LOGARITHMS**.

Now, what if the exponent changes in a power?

Then, the value of the power changes, too, of course.   How though?

Using a function, we can see how values change.
More specifically, how a value changes as other value changes.

So using a function, we can see how the value of a power changes as the value of its exponent changes.   What function then, is it?

It is called an exponential function.   What function is it though?

If an expression is a power, or has a power, it is called an exponential expression.

So an exponential function is an exponential expression, which can be a power, or can be made of powers.
For instance, an exponential function can be any of the expressions below:

$$2^x, \quad -3^x, \quad 3^x + 2^{x+1}, \quad 3^x - 2^{x+1} + 9, \quad 3^x - 5 \cdot 2^{x+1} + 1, \quad 2a^x - b^{x+1} + c, \text{ etc.}$$

And thus, of an exponential function, the expression has a power, where the base is usually a number or a constant, and the exponent is the input variable.

In short, an exponential function has a variable exponent.

So a function exponential is a function of an exponent.

And for instance, it can be any of the functions below:

$$y = f(x) = a^x \text{ for } x \text{ real}, \quad y = g(x) = 2^x + b \text{ for } x \geq 3, \quad \text{and } y = h(x) = 3^x + x + 1 \text{ for } x \geq 0.$$

In all the functions above, a^x, 2^x, and 3^x are powers, and in each, the exponent is the input variable.

In some exponential functions however, we can use an input variable as a base, too. And such a function can be $y = f(x) = x^x$. In high school math though, we don't use such functions.

- Note however, no matter what exponential function it may be, the base is positive, but is unequal to 1. The base cannot be ≤ 0, and cannot be 1, either.

Why not though, negative, 0, or 1?

We have: $0^2 = 0^3 = ... = 0$, and $1^2 = 1^{0.3} = ... = 1$.

So if the base is 0 or 1, there is not much of any significance in making such a function.

And if the base is negative, we cannot use many numbers as the exponent. Why not?

If the exponent is a fraction where the numerator is odd, but the denominator is even, and the base is negative, the function cannot be defined. That's because such a power cannot exist in the real number space. That is, such a power is not a real number.

For instance, $\sqrt{-2} = (-2)^{\frac{1}{2}}$ is not a real number, and we have infinitely many of such numbers even in such a tiny interval as $1 < x < 1.000000000000000000001$.
So we can hardly set up a domain if the base is negative.

And thus, using a constant as the base, we want to consider two cases, one is that the base is between 0 and 1, and the other is that the base is greater than 1.

So assuming $y = f(x) = a^x$, we want to consider the case where $0 < a < 1$, and the other case where $a > 1$. And the same is true, too, for a function as $y = g(x) = 2a^x + 3x^2 - x + 1$.

And in fact, the function behaves very differently depending on the case the base belongs to. The behavior in one case is directly opposite of the behavior in the other.

How then, do we make an exponential function?

Making a function, we define it, so making a function exponential, we define it, too. And defining it, we produce a description of a function. Then, the description is a function definition, called a definition of a function, too.

So a definition of an exponential function can be as follows: $y = g(x) = 2^x$ for $x > 3$.

And the function above is called **g**, which is therefore, the name of the function, and is defined only if $x > 3$, which is thus, the domain, so **x** cannot get any value bigger than 3.

And defining in general, such a function as above, we can define it the way below:

$y = f(x) = a^x$ where $a > 0$, but $a \neq 1$.

That's because the base has to be positive, but is not equal to 1.
So just given $y = f(x) = a^x$ only, we need to assume that $a > 0$, but $a \neq 1$.

What then, about the domain?

If not specified with a function, the domain is assumed to be the largest set of numbers the function can hold for. The function f can hold for all real numbers, that is, **x** can be all real numbers, so the domain is a set of all real numbers.   What then, is the range?

We know the base **a** is positive, and not equal to 1, so **a** is positive anyway, and thus, every number a^x produces is positive, that is, every output is positive. And we know the range is the set of all the outputs, and **x** takes all real numbers.   What then, is the range?

The range is a set of all *positive* numbers.

So assuming **R** is a set of all real numbers, and **r** is a set of all *positive* numbers, we can set: $f: R \longrightarrow r$. Specifically, we can set: $f: x \longrightarrow a^x$ where $x \in R$, $a > 0$, $a \neq 1$.

So it is said that the exponential function f is a function from **R** to **r**.

And of course, that's not the only exponential function we can have.

For instance, another function exponential can be: $y = g(x) = -2c^x + 3x - 1$ for $x > 0$.

And in that case, too, it is assume that the base $c > 0$, and $\neq 1$.

And the same is true for these, too: $y = p(x) = m^{3x+1} + x + 1$, and $y = q(x) = m^x n^{2x}$.

So in those cases, too, it is assumed that $m > 0$, and $\neq 1$, and that $n > 0$, and $\neq 1$.

- And we know some functions can have their inverses.

So what about the inverse of an exponential function?

If a function has its inverse, the function is invertible, and thus, is one-to-one.
So if we want to see if a function is invertible, that is, if it can have its inverse, we want to check to see if the function is one-to-one.

In the case of a function one-to-one, the same output cannot be produced more than once.

So showing a function f is one-to-one, we want to show: $f(m) \neq f(n)$, assuming of course, $m \neq n$ where m and n are constant, and belong to the domain.

So let's now check to see if the function f below is one-to-one.

$y = f(x) = a^x$ where $a > 0$ but $a \neq 1$.

To begin with, we can get: $\dfrac{f(m)}{f(n)} = \dfrac{a^m}{a^n} = a^{m-n}$. The domain is a set of all real numbers.

Assuming thus, $m \neq n$ where m and n are constant and real, we get: $a^{m-n} \neq a^0 = 1$.

So we get: $\dfrac{f(m)}{f(n)} \neq 1$, and thus, we get: $f(m) \neq f(n)$. So f is one-to-one.

And thus, f is invertible, so it has its inverse. What then, is the inverse?

We know that of the exponential function f, each output is a value of a power, for instance, 9 is the value of 3^2, and each input is an exponent.

For instance, assuming $a = 3$ in a^x, we get: $y = f(x) = 3^x$, so we get: $f(2) = 3^2 = 9$.

So in short, of the function f, each output is a power, and each input is an exponent.

What then, is each output of the inverse?

It is an exponent. And each input of the inverse is (the value of) a power.

That's because outputs of the inverse are inputs of the original function f, and outputs of the inverse are inputs of the original function f.

And we have a function, of which each input is a number that is the value of a power, and each output is an exponent.

In short, of such a function, each input is a power, and each output is an exponent.

And such a function is called a logarithmic function, called a log function, for short.

• So the inverse of the exponential function f is a log function.

What then, is the domain of the log function above?

We know that the log function is the inverse of the exponential function f, and also that the domain of an inverse is the range of the original function, that is, the function we take the inverse of.

So the domain of the log function is the range of the exponential function f.

And we know that the range of the exponential function f is a set of all positive numbers.

So the domain of the log function is a set of all positive numbers.

What then, about the range of the log function?

We know that the domain of the exponential function f is a set of all real numbers.

So the range of the log function is a set of all real numbers.

Thus in short, the domain of the log function is a set of all positive numbers, and the range is a set of all real numbers.

And a log function is a function, too. So making a log function, we define it, too.
How then, do we define a log function?
In other words, how do we make a log function definition?

The log function stated above is <u>the inverse</u> of the exponential function f.

So beginning with <u>the definition for inverse functions</u>, we have: $y = f(x) \Leftrightarrow x = f^{-1}(y)$.

Next, we have: $y = f(x) = a^x$ where $a > 0$, but $a \neq 1$.   And the range of f is: $y > 0$.

And next, <u>the definition for logs</u> is as follows: $A = b^c \Leftrightarrow c = \log_b A$ where $b > 0$, but $\neq 1$.

So using <u>both the two definitions above</u>, we can put together the exponential function f and the inverse of f, that is, the log function the way as follows:

$$y = f(x) = a^x \Leftrightarrow x = f^{-1}(y) = \log_a y \text{ where } x \text{ is real, } y > 0, \text{ and } a > 0, \text{ but } a \neq 1.$$

What is f^{-1}, though?

It is the name of the log function, which is: $x = f^{-1}(y) = \log_a y$.

The name of the log function is however, a bit bulky to carry around with.

And we can name a function differently if consistency is kept. So we can rename the log function above, and for instance, we can call it g, and can set: $x = g(y) = \log_a y$ for $y > 0$.

We don't usually though, take the equality above as a log function definition.
It is quite conventional that we use x as the input variable, and use y as the output variable. That is, we usually put a function in the x-y system, and not y-x system.

So assuming the log function is called g, and putting it in the x-y system, we can put the log function g the way as follows:

$y = g(x) = \log_a x$ for $x > 0$.

Then, of the log function g above, each $\underline{x\text{-value}}$ is an input, and is the value of $\underline{\text{a power}}$ a^y for each $\underline{y\text{-value}}$, which is an output, and is $\underline{\text{an exponent}}$.

For instance, if $a = 2$ and $x = 8$, we get: $y = 3$, because $8 = 2^3$. So $3 = g(8) = \log_2 8$.

And we can put all the ideas above formally (that is, symbolically) the way below:

First, assuming R is a set of all real numbers, and r is a set of all positive numbers, we can set: $f: R \longrightarrow r$. So the exponential function f is a function from R to r.

And specifically, we can set: $f: x \longrightarrow a^x$ where $x \in R$, $a > 0$, and $\neq 1$.

And usually, we just put it this way: $y = f(x) = a^x$.

And next, g is the inverse of f, and we know the domain of an inverse is the range of the original function, and the range of the inverse is the domain of the original.

So we can set: $g: r \longrightarrow R$.   And thus, the log function g is a function from r to R.

And specifically, we can set: $g: x \longrightarrow \log_a x$ where $x \in r$, $a > 0$, and $\neq 1$.

And usually, we just put it this way: $y = g(x) = \log_a x$.

And just given: $y = f(x) = a^x$, we need to assume that x is real, $a > 0$, and $a \neq 1$.

And also, just given: $y = g(x) = \log_a x$, we want to assume that $x > 0$, $a > 0$, and $a \neq 1$.

And we want to keep in mind that a log function is an inverse of an exponential function, and vice versa. That is, both are the inverse of each other. And in short:

Each input of an exponential function is an exponent, and the output is a power.

Each input of a log function is a power, and the output is an exponent.

How do we call then, x in $\log_a x$?

It is called an antilogarithm, often just called an antilog, for short.

What then, is $\log_a x$?

It is in fact, an exponent, since if we have: $y = g(x) = \log_a x$, we get: $x = a^y$, where y is an exponent.   So $(\log_a x)$ itself is an exponent.

Specifically thus, we can put an exponent in a structured form. An exponent in such a form above can let us see specifics on itself.

For instance, specifying 81, we can use 2 as the exponent using 9 as the base, and also, we can use 4 as the exponent using 3 as the base.   So we can have: $81 = 9^2 = 3^4$.

Specifying a number therefore, we can use a different exponent using a different base. And thus, a log shows what an exponent is about. We have: $2 = \log_9 81$, and $4 = \log_3 81$.

And the same is true, too, for specifying an exponent.

So specifying an exponent, we can use a different number using a different base. For instance, we can have: $8 = 2^3$, $27 = 3^3$, $64 = 4^3$, etc.

And thus, using logarithms, we can put the exponent 3 in such many ways as below:

$$3 = \log_2 8 = \log_3 27 = \log_4 64 = \ldots$$

So each of $\log_2 8$, $\log_3 27$, and $\log_4 64$ is a number, is an exponent, and is 3.

And thus, $\log_a x$ is in fact, an exponent.

So for another instance, we can put 16 this way, too: $16 = 2^{\log_2 16}$ since $4 = \log_2 16$, which is thus, an exponent. We can therefore, indicate x in $\log_a x$ this way, too: $x = a^{\log_a x}$.

Now, let's take a look at what's happening in the exponential function $y = f(x) = a^x$ as the input changes. That is, we are going to keep track of the y-value as the x-value changes. Then, we can see how the exponential function f behaves.

We have two cases though. In one case, the base a is positive but is less than 1, and in the other, the base a is greater than 1.

So beginning with the case where $a > 1$, we can see y gets increased as x gets increased.

That is, as the exponent increases, the power increases also if the base is bigger than 1.

So for instance, we have: $\ldots 2^{-3} < 2^{-2} < 2^{-1} < 2^0 = 1 < 2^1 < 2^2 < 2^3 < 2^4$, and so forth.

How large then, the value of y can be?

It is the value of a power, and will be infinitely large if the exponent is infinitely large, that is, if the exponent x goes infinity, because the base is bigger than 1.

And thus, if the base a is bigger than 1, the power a^x keeps increasing as the exponent x increases, and the power goes infinity when the exponent goes infinity.

If however, we have: $0 < a < 1$, we can see that y gets decreased as x gets increased.

That is, as the exponent increases, the power decreases if the base is between 0 and 1. So for instance, we have:

... $0.1^{-2} = 10^2 > 0.1^{-1} = 10 > 0.1^0 = 1 > 0.1^1 > 0.1^2 > 0.1^3 > 0.1^4$, and so forth.

How small then, the value of y can be? Can it be 0 or negative?

We know: $y > 0$, so y cannot be negative, and it cannot be 0, either, but can be infinitely small being positive, though.   How small then, can it be anyway?

It will be infinitesimal if the exponent is infinitely large, that is, if in a^x, the exponent x goes infinity. An infinitesimal is as good as 0, but is not 0 itself.

And thus, if the base a is bigger than 1, as the exponent increases, the power a^x keeps decreasing, and as the exponent x grows bigger, the power a^x approaches 0, then eventually, will be infinitesimal when the exponent x goes infinity.

So we can put in a graph, the curve of the exponential function f the way as follows:

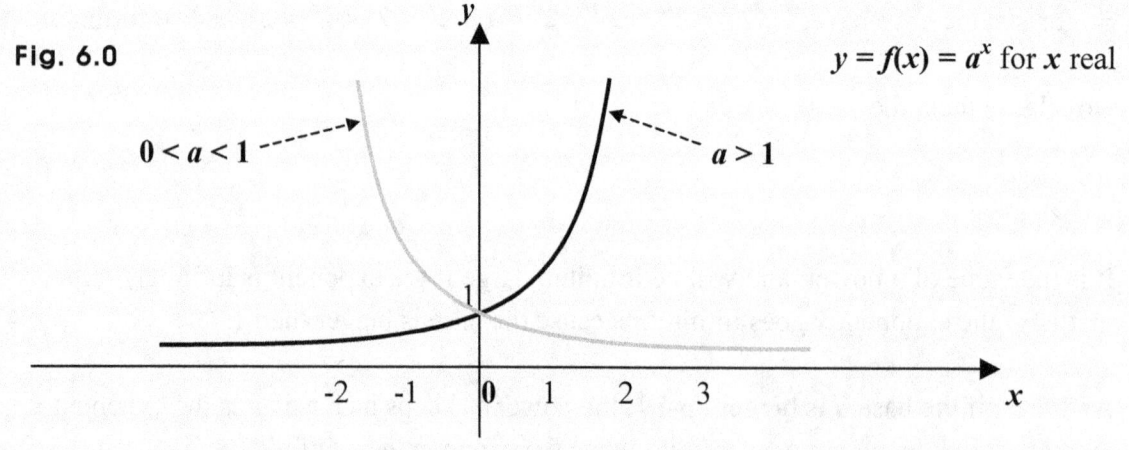

Fig. 6.0

$y = f(x) = a^x$ for x real

$0 < a < 1$

$a > 1$

Notice that the two curves meet at the point **(0, 1)**.

Now, let's next, take a look at what's happening in the log function $y = g(x) = \log_a x$ as the input changes. That is, we are going to keep track of the value of y as x changes its value. Then, we can see what the curve of the log function g looks like.

As in the case of the exponential function, we have two cases. In one case, the base a is positive but is less than 1, and in the other, the base a is greater than 1.

So beginning with, the case where $a > 1$, we can see y gets increased as x gets increased.

That is, as the antilog increases, the exponent increases, too, if the base is bigger than 1.

So for instance, we have:

$$\ldots \ \log_2 \tfrac{1}{8} = -3 < \log_2 \tfrac{1}{4} = -2 < \log_2 \tfrac{1}{2} = -1 < \log_2 1 = 0 < \log_2 2 = 1 < \log_2 4 = 2 \ldots$$

That's because we have: $\ldots 2^{-3} < 2^{-2} < 2^{-1} < 2^0 = 1 < 2^1 < 2^2 < 2^3 < 2^4$, and so forth.

How large then, the exponent y, that is, the value of $(\log_a x)$ can be?

It will be infinitely large if the antilog x is infinitely large, that is, goes infinity.

And thus, if the base a is bigger than 1, the exponent y keeps increasing as the antilog x increases, and the exponent $(y = \log_a x)$ goes infinity when the antilog x goes infinity.

If however, $0 < a < 1$, we can see y decreases as x increases.

That is, as the antilog increases, the exponent decreases if the base is between 0 and 1. So for instance, we have:

$$\ldots \ \log_{0.1} \tfrac{1}{100} = 2 > \log_{0.1} \tfrac{1}{10} = 1 > \log_{0.1} 1 = 0 > \log_{0.1} 10 = -1 > \log_{0.1} 100 = -2 \ \ldots$$

That's because we have:

$$\ldots 0.1^{-2} = 10^2 > 0.1^{-1} = 10 > 0.1^0 = 1 > 0.1^1 > 0.1^2 > 0.1^3 > 0.1^4, \text{ and so on.}$$

How small then, the exponent y can be? Can it go negative infinity?

Yes, it will go negative infinity if the antilog is infinitely large, that is, if x goes infinity.

And thus, if the base is between 0 and 1, the exponent (the y-value) keeps decreasing as the antilog (the x-value) increases, and the exponent goes negative infinity when the antilog goes infinity.

So putting in a graph, the curve of the log function g, we can put it the way below.

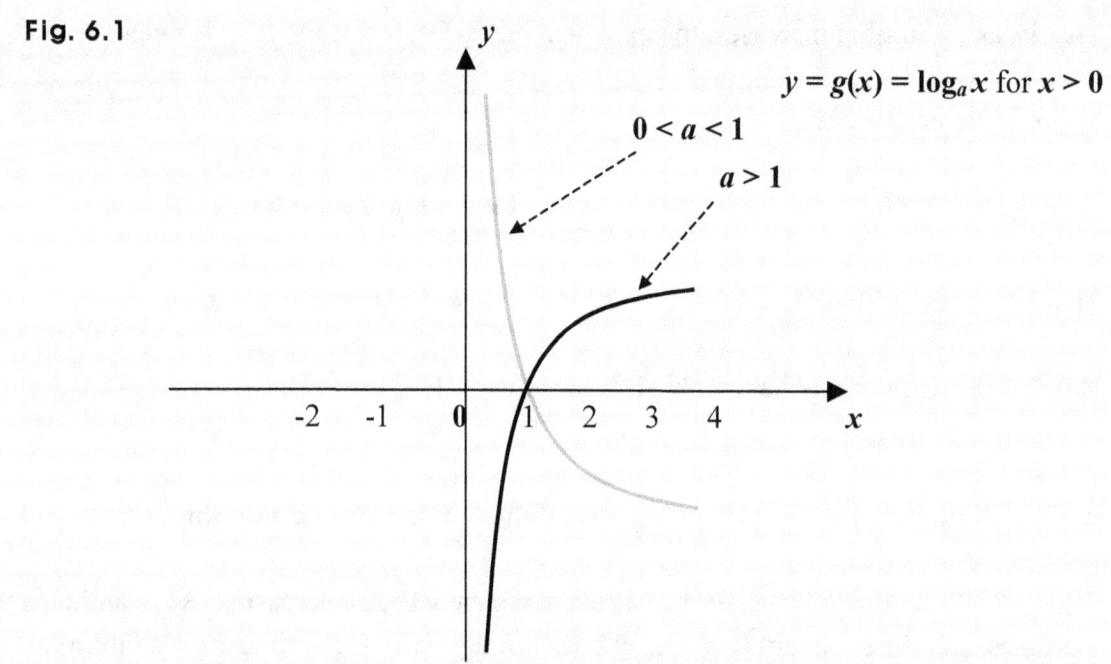

Fig. 6.1

$y = g(x) = \log_a x$ for $x > 0$

$0 < a < 1$

$a > 1$

We can notice that the two curves meet at the point **(1, 0)**.

And note that if $a > 1$, the y-value keeps increasing, then goes infinity as the x-value keeps increasing, then goes infinity, and that if $0 < a < 1$, the y-value keeps decreasing, then goes negative infinity as the x-value keeps increasing, then goes infinity.

Now, we have: $y = f(x) = a^x$ for x real, and $y = g(x) = \log_a x$ for x > 0.

And we know that the log function g is the inverse of the exponential function f.

So the curves of both functions are symmetric about the line $y = x$.

And thus, putting in a graph both curves of the two functions f and g, we get:

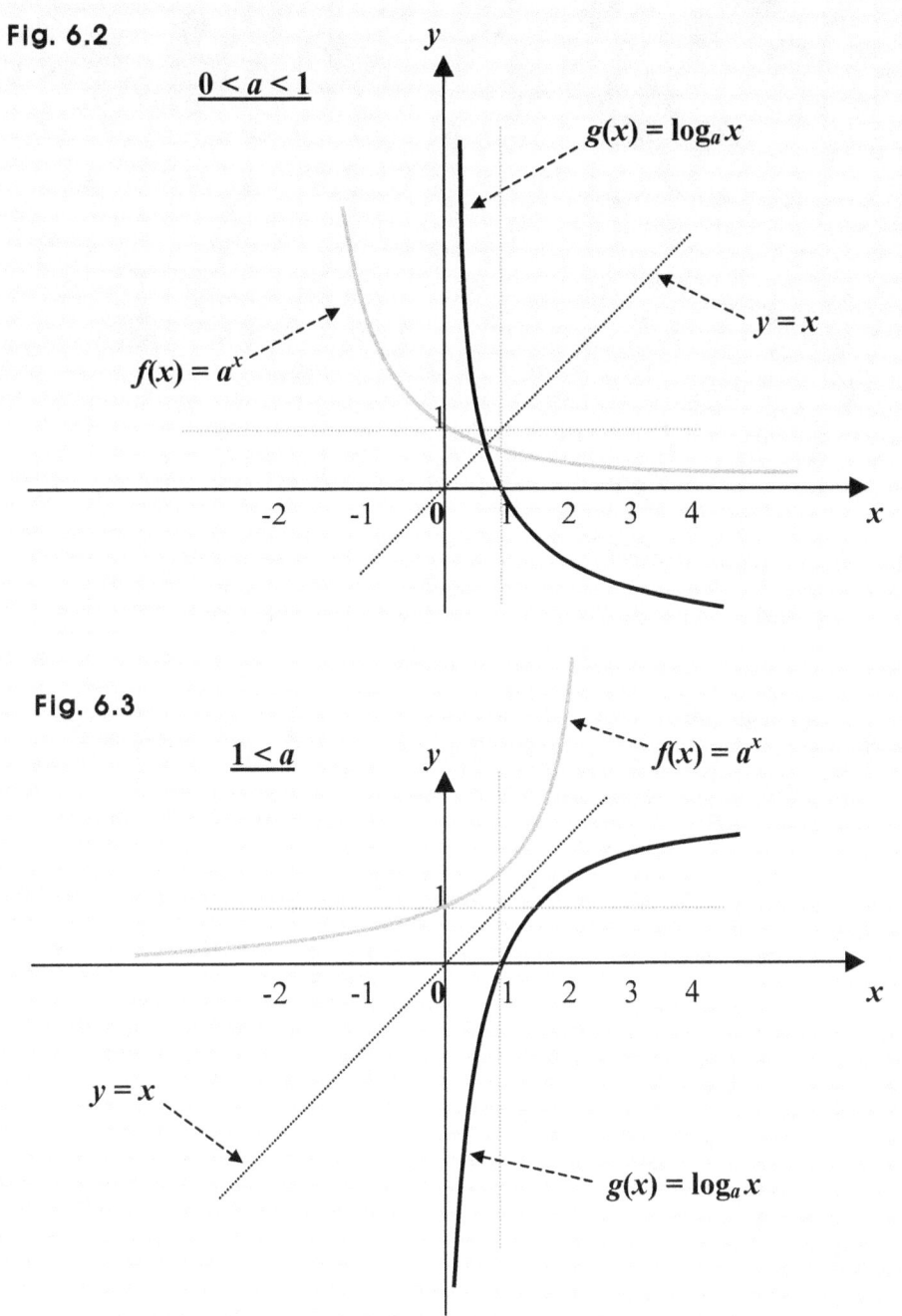

Fig. 6.2

$0 < a < 1$

$g(x) = \log_a x$

$y = x$

$f(x) = a^x$

Fig. 6.3

$1 < a$

$f(x) = a^x$

$y = x$

$g(x) = \log_a x$

And in this case, the line $y = 0$ (i.e., the x-axis) is called the asymptote for $y = a^x$. Specifically, it is called a horizontal asymptote.

Also, the line $x = 0$ (i.e., the y-axis) is called the asymptote for $y = \log_a x$, and specifically, it is called a vertical asymptote.

Examples 1 in Exponential & Log Functions

In this set of examples, we construct the graph of a function, which is exponential. Constructing it, we put in the graph, the curve of the function. That is, we get the curve of the function.

And a function is said to make or produce its curve, too, because it generates each and every point in the curve.

We are going to cover though, not just one graph, but many graphs.
So we will get to work with many different functions, which are quite similar though.

And equations have their curves, too.
So we will get to see how to make and work with curves of some functions similar, and curves of some equations, too. And thus, this example will take quite a few pages.

That's because graphing matters. And so does algebra, of course.

Graphing, you get the curve of a function or equation, and can actually see what the function or equation is doing, and thus, can see better what you can do about it.
So what?

So you can get to the solution easily and quickly, together with, of course, your good algebra, because it actually connects the problem to the solution.

And the example is as follows:

Construct the graph of a function as follows: $y = f(x) = 2^{|x|}$ for x real.

Suggestions or Solutions
To the Problem in the Example Given

Construct the graph of $y = f(x) = 2^{|x|}$ for x real.

Putting a function in a graph, we get the curve of it, and can actually see how it behaves, so doing problems with functions, we can see better the solutions' whereabouts.

Now, the function given is an exponential function, of which the domain is a set of all real numbers. In its expression however, absolute sign is applied to the input variable, which is x, of course. So what does the curve look like?

The curve is symmetric about the y-axis. How come?

If the domain is symmetric about the origin, and absolute sign is applied to every input variable, the curve is symmetric about the y-axis.
What do we mean by the domain symmetric about the origin, though?

Suppose for instance, a is constant, a function is defined for $|x| \le a$, that is, $-a \le x \le a$, another is defined for $|x| < a$, that is, $-a < x < a$, and another is defined for x real, that is, a set of all real numbers. Then, each of all the domains is symmetric about the origin.

And if absolute sign is applied to every input variable, and the input is negative, the output is the same as the one for an input positive and equal in magnitude. So for instance, we get: $f(-1) = f(1)$, and the same is true, too, for all the other inputs negative.

And thus, the curve is symmetric about the y-axis.

So let see now, how the curve is symmetric about the y-axis.

Assuming first, $y = g(x) = a^x$ for x real, and putting it in a graph, we will get:

Fig. 0

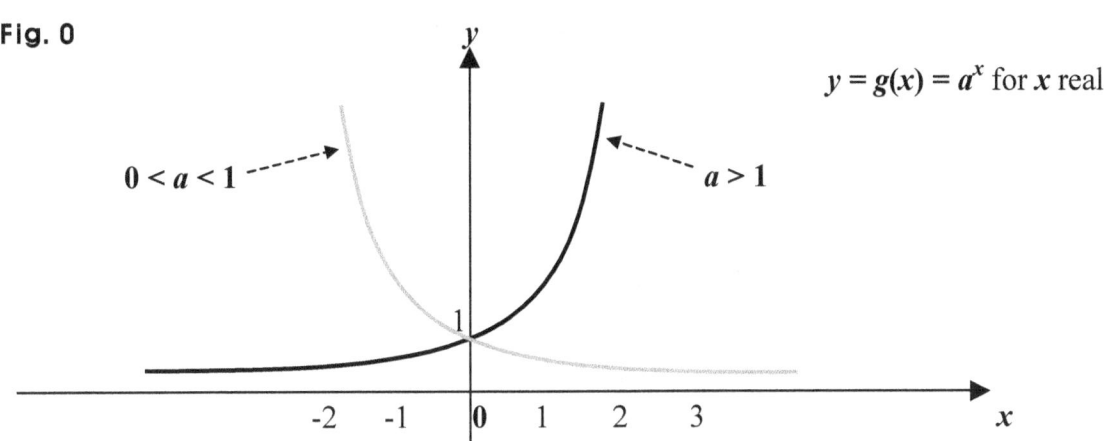

$y = g(x) = a^x$ for x real

$0 < a < 1$

$a > 1$

Both curves gray and black pass through the point **(0, 1)**, of course.

And for instance, if the base a is 2, the curve of the function g will be similar to the curve in black above, and thus, will be as follows.

Fig. 1

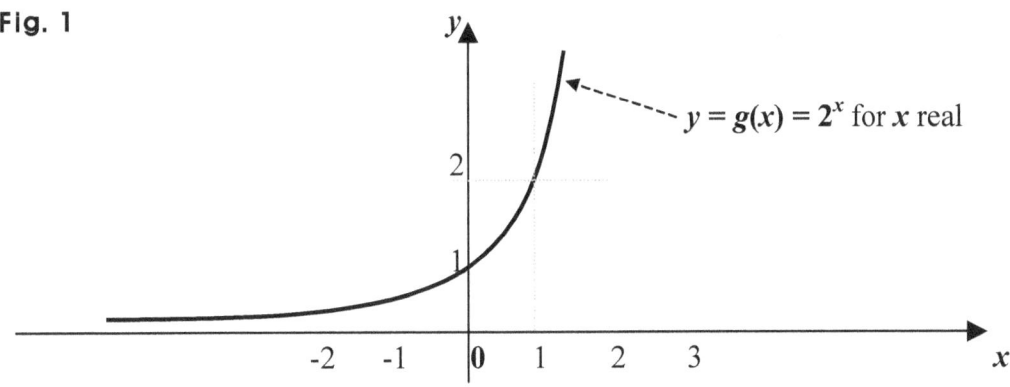

$y = g(x) = 2^x$ for x real

So in the graph above, when $x \geq 0$, the amount of the curve of g is as below.

Fig. 2

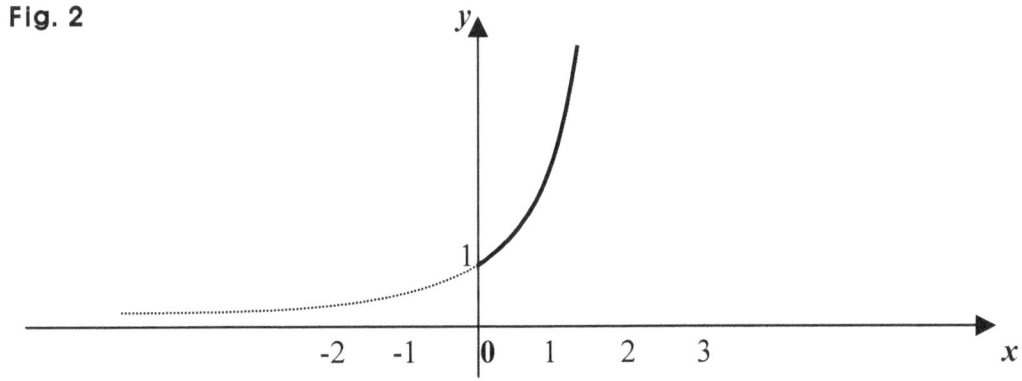

And next, putting in a graph, another function $y = h(x) = 2^{-x}$ for x real, we get:

Fig. 3

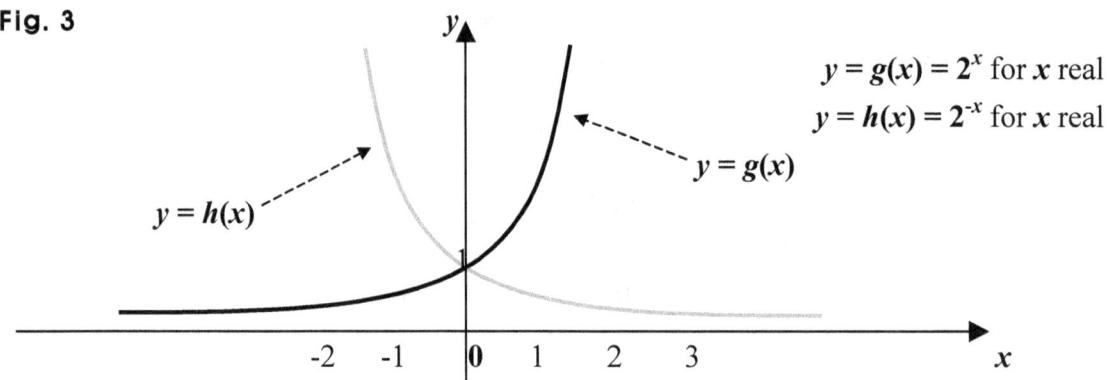

$y = g(x) = 2^x$ for x real
$y = h(x) = 2^{-x}$ for x real

How come?

We have: $2^{-x} = (2^{-1})^x = (\frac{1}{2})^x$.

So the graph of h is like the graph of the function $y = g(x) = a^x$ where $0 < a < 1$.
And in fact, the curve of h is symmetric about the y-axis to the curve of $y = g(x) = 2^x$.
How come?

Suppose in h, the x-value changes from 0 to -2, for instance.

How then, does the value of $(-x)$ change?

It changes from 0 to 2, of course. And we have: $y = h(x) = 2^{-x}$. So when x changes from 0 to -2, the value of $h(x)$ is the same as the value of $g(x)$ when x change from 0 to 2.

Thus for instance, $h(0) = g(0)$, $h(-1) = g(1)$, and $h(-2) = g(2)$,

So when x changes from 0 to -2, the curve itself made by $h(x)$ is exactly the same as the curve made by $g(x)$ when x changes from 0 to 2, but is proceeding in the opposite direction. And thus, both curves are symmetric about the y-axis.

Fig. 4

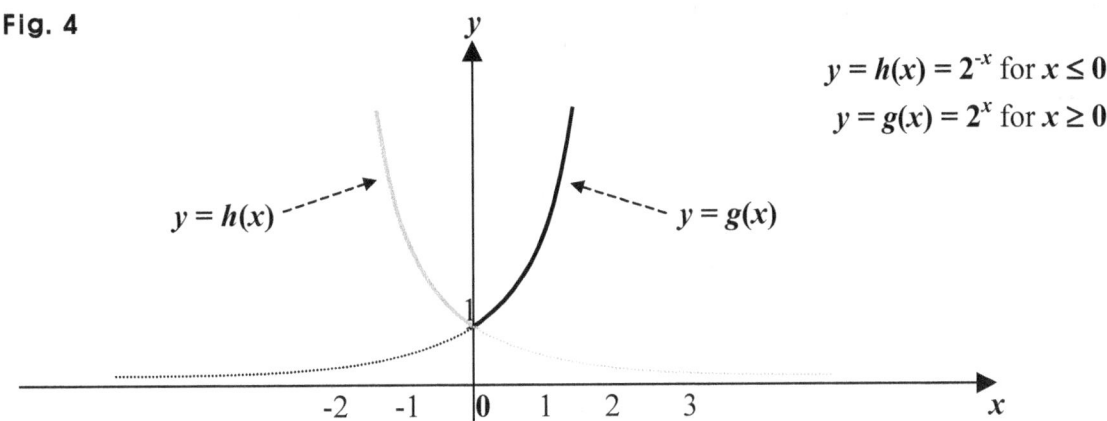

$$y = h(x) = 2^{-x} \text{ for } x \le 0$$
$$y = g(x) = 2^{x} \text{ for } x \ge 0$$

Now, in this problem, we have: $y = f(x) = 2^{|x|}$ for x real.

To begin with, we get: $f(x) = f(-x)$, because $2^{|x|} = 2^{|-x|}$. What then, is the value of $f(x)$ equal to when the x-value changes from 0 to -2, for instance?

It is equal to the value of $f(x)$ when the x-value changes from 0 to 2.

Thus, for instance, we have: $f(-1) = 2^{|-1|} = 2^1 = 2$, and $f(1) = 2^{|1|} = 2$, so we get:

$f(-1) = f(1)$. And the same is true, too, for all the other values of x.

So when x changes from 0 to -2, the curve itself made by f is exactly the same as the curve made by f when x changes from 0 to 2, but is proceeding in the opposite direction.

Thus, both curves are symmetric about the y-axis. And we can put it the way below, too:

When x changes from -2 to 0, the curve itself made by f is exactly the same as the curve made by f when x changes from 0 to 2, but is going downward. Thus again, both curves are symmetric about the y-axis.

So when x changes from negative infinity to 0, the curve itself made by f is exactly the same as the curve made by f when x change from 0 to infinity, but is going downward.

And thus, both curves are symmetric about the y-axis.

So the entire curve of f is symmetric about the y-axis, and is as follows:

Fig. 5

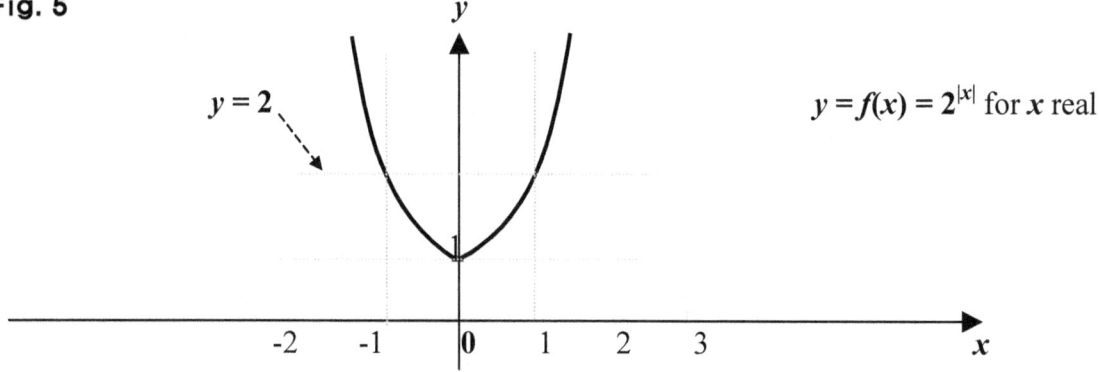

$y = f(x) = 2^{|x|}$ for x real

And quite often, we put the function above this way, too: $y = f(|x|) = 2^{|x|}$ for x real.

So for instance, setting $y = p(|x|)$, we apply absolute sign to every x in the expression of p, and thus, if $y = p(x) = 2^{x+1} + 2x + 1$, we get: $y = p(|x|) = 2^{|x|+1} + 2|x| + 1$.

Suppose next, $y = g(x) = 2^{|x|}$ for $x \geq -1$.

Then, the domain is not symmetric about the origin, so the curve of g is not symmetric about the y-axis, and we get:

Fig. 6

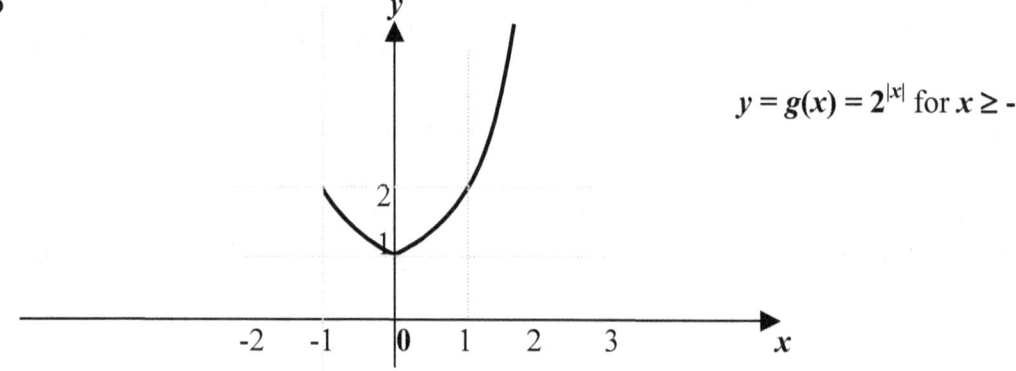

$y = g(x) = 2^{|x|}$ for $x \geq -1$

How then, about the curve of $y = g(x) = 2^{x-1}$?

We can readily get it shifting by 1 the curve of a function $y = f(x) = 2^x$ in the direction of the x-axis.   How come?

Suppose in g, the x-value changes from 1 to 3, for instance.

How then, does the value of $(x - 1)$ change?

It changes from 0 to 2, of course. And we have: $y = f(x) = 2^x$. So when x changes from 0 to 2, the value of $f(x)$ is the same as the value of $g(x)$ when x changes from 1 to 3.

Thus, for instance, $f(0) = g(1)$, $f(1) = g(2)$, and $f(2) = g(3)$.

So when x changes from 1 to 3, the curve made by $g(x)$ is exactly the same as the curve made by $f(x)$ when x changes from 0 to 2, and will behave in the same manner.

And thus, shifting by 1 the entire curve of f in the direction of the x-axis, we can get the entire curve of g.

Fig. 7

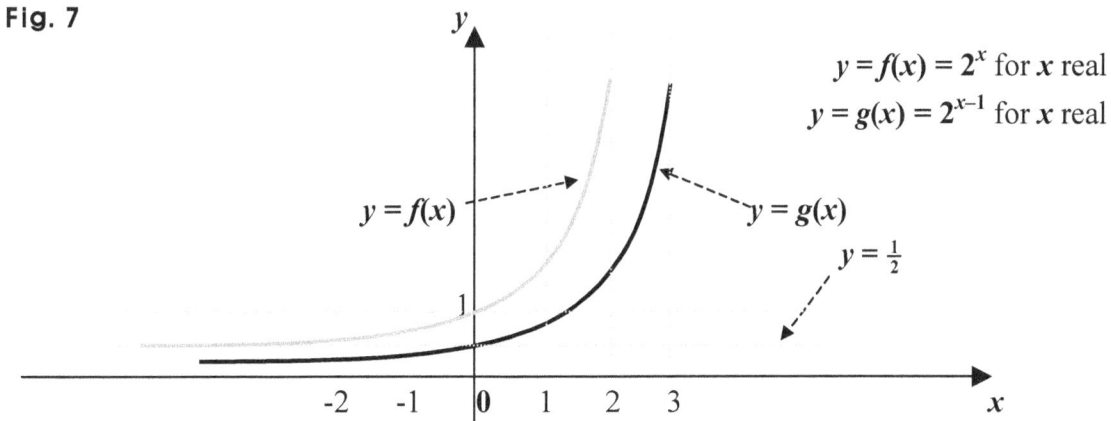

$y = f(x) = 2^x$ for x real
$y = g(x) = 2^{x-1}$ for x real

$y = f(x)$

$y = g(x)$

$y = \frac{1}{2}$

-2   -1   0   1   2   3 x

And the same is true, too, for the curve of a function $y = h(x) = 2^{|x-1|}$.

That is, shifting by 1 the curve of a function $y = f(x) = 2^{|x|}$ in the direction of the x-axis, we can readily get the curve of $y = h(x) = 2^{|x-1|}$.

Fig. 8

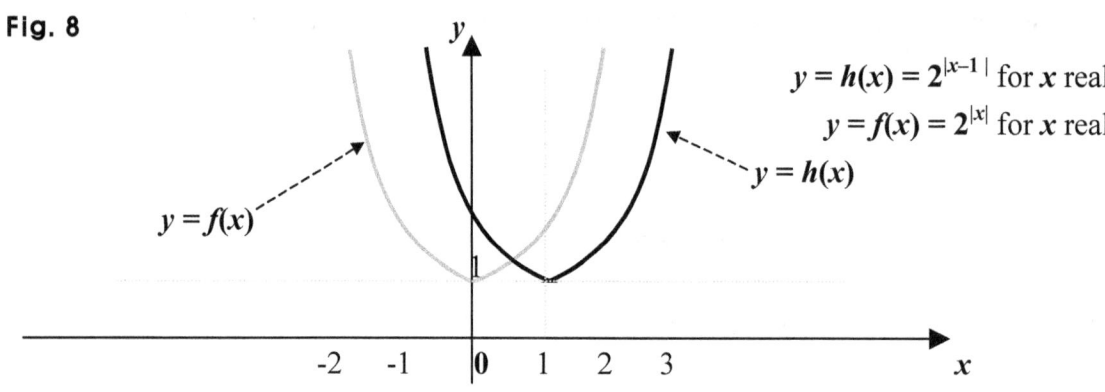

$y = h(x) = 2^{|x-1|}$ for x real

$y = f(x) = 2^{|x|}$ for x real

$y = h(x)$

$y = f(x)$

-2   -1   0   1   2   3

x

How then, about the curve of this function: $y = g(x) = 2^{x+1}$?

We can readily get it shifting by -1 the curve of a function $y = f(x) = 2^x$ in the direction of the x-axis.   How come?

Suppose in g, the x-value changes from -1 to 2, for instance.

How then, does the value of $(x + 1)$ change?

It changes from 0 to 3, of course. And we have: $y = f(x) = 2^x$.

So when x changes from 0 to 3, the value of $f(x)$ is the same as the value of $g(x)$ when x changes from -1 to 2.

Thus, for instance, $f(0) = g(-1)$, $f(1) = g(0)$, and $f(3) = g(2)$.

So when x changes from -1 to 2, the curve made by $g(x)$ is exactly the same as the curve made by $f(x)$ when x changes from 0 to 3, and behaves the same manner.

And thus, shifting by -1 the entire curve of f in the direction of the x-axis, we can get the entire curve of g.

Fig. 9

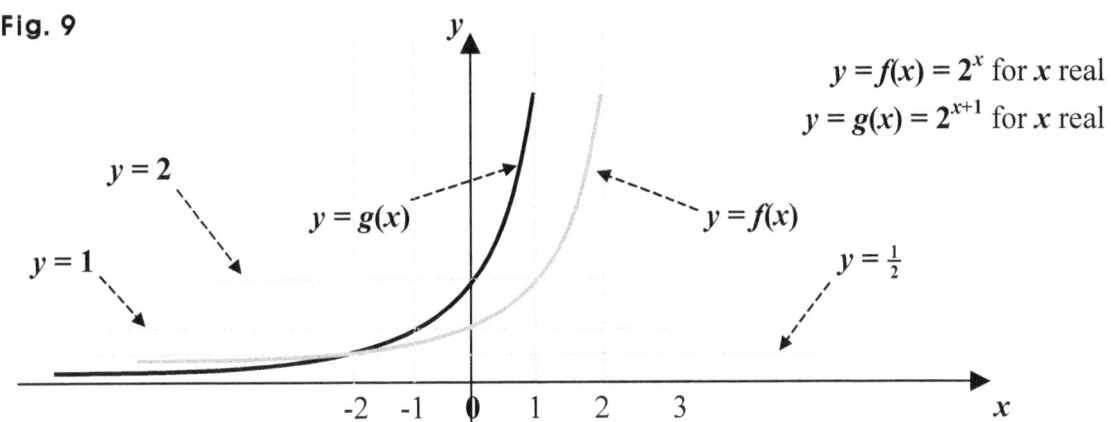

$$y = f(x) = 2^x \text{ for } x \text{ real}$$
$$y = g(x) = 2^{x+1} \text{ for } x \text{ real}$$

And the same is true, too, for the curve of another function $y = h(x) = 2^{|x+1|}$.

That is, shifting by -1 the curve of a function $y = f(x) = 2^{|x|}$ in the direction of the x-axis, we can readily get the curve of $y = h(x) = 2^{|x+1|}$.

Fig. A

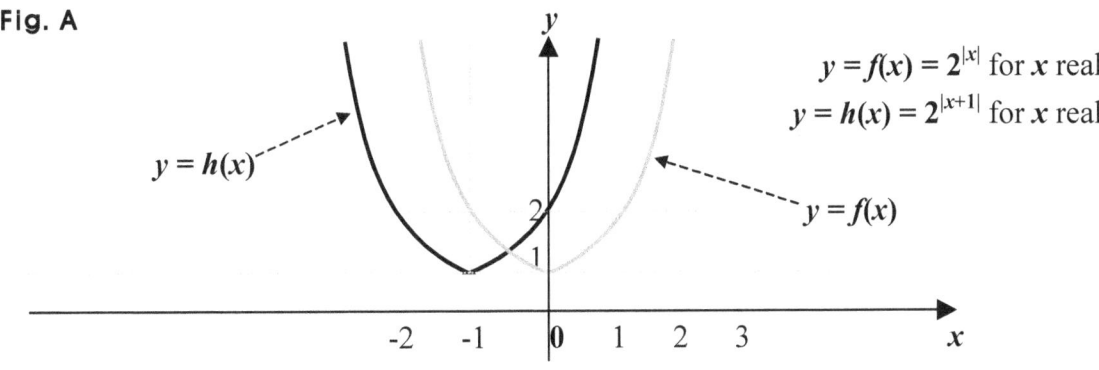

$$y = f(x) = 2^{|x|} \text{ for } x \text{ real}$$
$$y = h(x) = 2^{|x+1|} \text{ for } x \text{ real}$$

Next, how about the curve of another function $y = g(x) = 2^x + 1$?

We can readily get it shifting by 1 the curve of a function $y = f(x) = 2^x$ in the direction of the y-axis.   How come?

Every time we add 1 to an output for a particular input of the function f, the sum is the output for the same particular input of the function g.

That is, we have: $g(x) = f(x) + 1$, because $g(x) = 2^x + 1$ and $f(x) = 2^x$.

And thus, shifting by 1 the curve of f in the direction of the y-axis, we can get the curve of g.

Fig. B

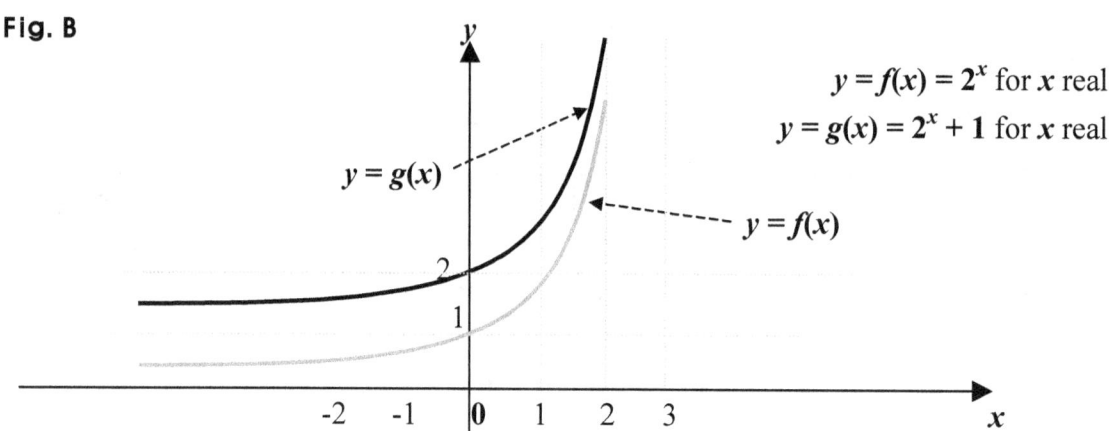

And the same is true, too, for the curve of another function $y = h(x) = 2^{|x|} + 1$.

That is, shifting by 1 the curve of a function $y = f(x) = 2^{|x|}$ in the direction of the y-axis, we can readily get the curve of $y = h(x) = 2^{|x|} + 1$.

Fig. C

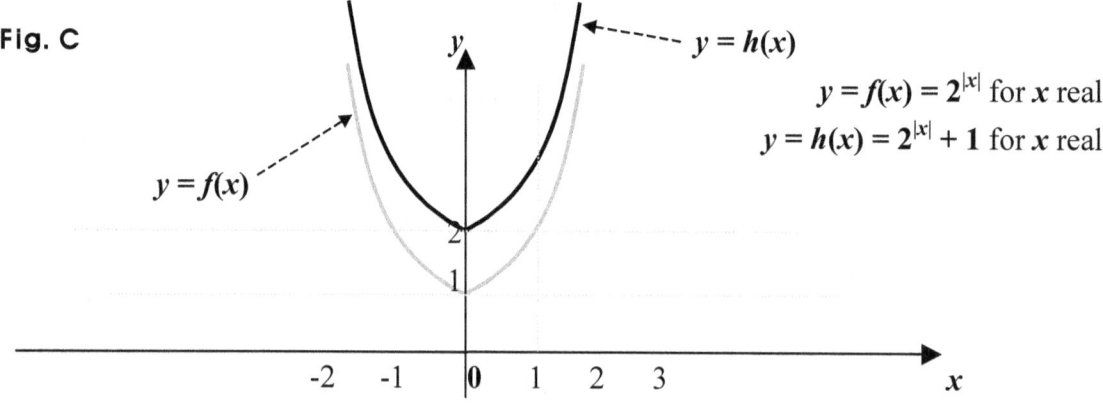

How then, about the curve of this function: $y = g(x) = 2^x - 1$?

We can readily get it shifting by -1 the curve of a function $y = f(x) = 2^x$ in the direction of the y-axis.   How come?

Every time we subtract 1 from an output for a particular input of the function f, the difference is the output for the same particular input of the function g.

That is, we have: $g(x) = f(x) - 1$, because $g(x) = 2^x - 1$ and $f(x) = 2^x$. And thus, shifting by -1 the curve of f in the direction of the y-axis, we can get the curve of g.

Fig. D

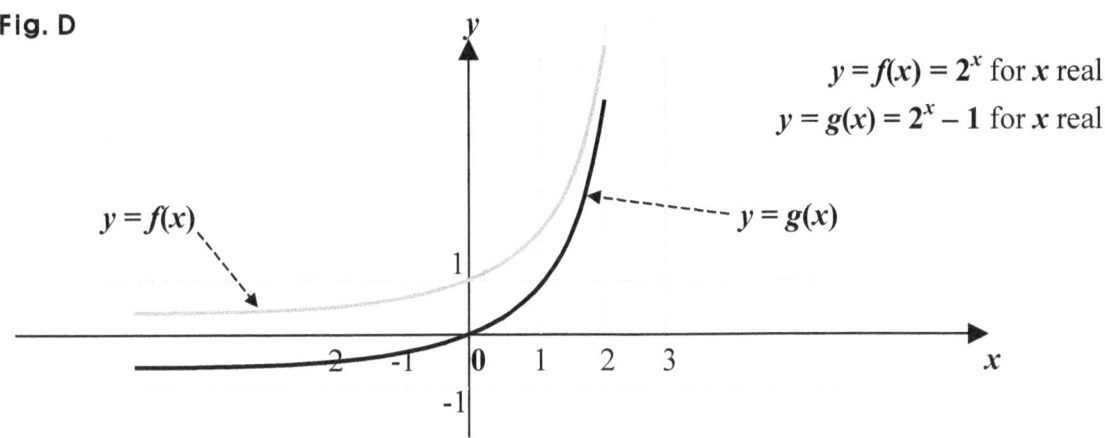

$$y = f(x) = 2^x \text{ for } x \text{ real}$$
$$y = g(x) = 2^x - 1 \text{ for } x \text{ real}$$

And the same is true, too, for the curve of another function $y = h(x) = 2^{|x|} - 1$.

That is, shifting by -1 the curve of a function $y = f(x) = 2^{|x|}$ in the direction of the y-axis, we can readily get the curve of $y = h(x) = 2^{|x|} - 1$.

Fig. E

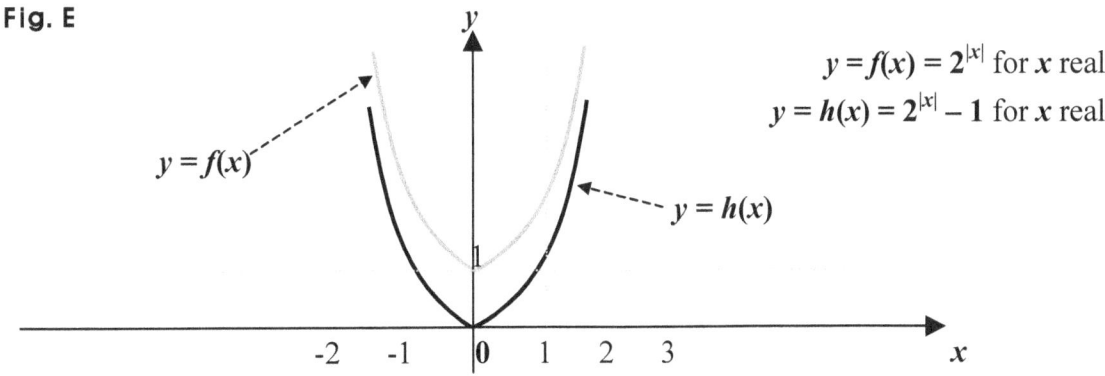

$$y = f(x) = 2^{|x|} \text{ for } x \text{ real}$$
$$y = h(x) = 2^{|x|} - 1 \text{ for } x \text{ real}$$

Next, how about the curve of another function $y = u(x) = 2^{x-1} + 1$?

We can readily get it shifting by 1 the curve of a function $y = v(x) = 2^{x-1}$ in the direction of the y-axis.   How come?

Every time we add 1 to an output for a particular input of the function v, the sum is the output for the same particular input of the function u.

That is, we have: $u(x) = v(x) + 1$, because $u(x) = 2^{x-1} + 1$ and $v(x) = 2^{x-1}$.
And thus, shifting by 1 the curve of v in the direction of the y-axis, we can get the curve of u.
And also, we know we can get the curve of $v(x) = 2^{x-1}$ shifting by 1 the curve of a function $y = f(x) = 2^x$ in the direction of the x-axis

So we can readily get the curve of $u(x) = 2^{x-1} + 1$ shifting by 1 the curve of $y = f(x) = 2^x$ in the direction of the x-axis, and then, shifting by 1 in the direction of the y-axis.

Fig. F

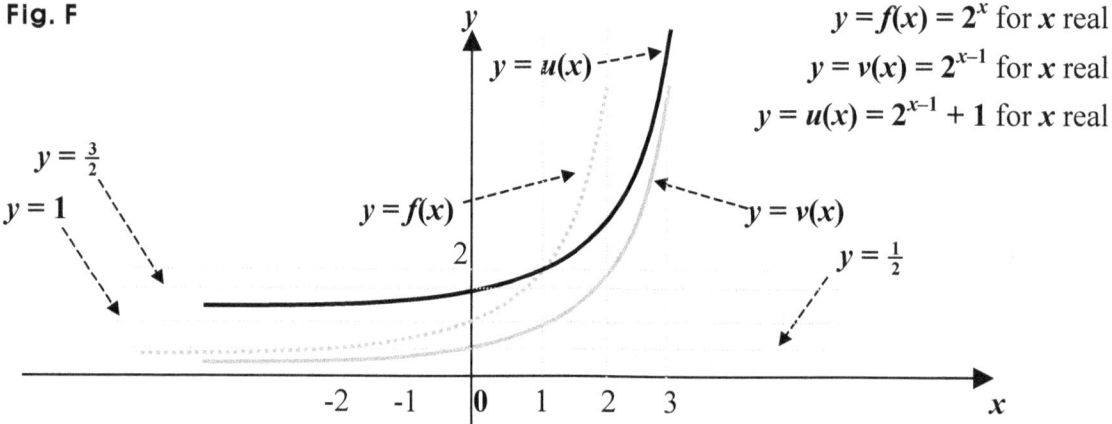

$y = f(x) = 2^x$ for x real
$y = v(x) = 2^{x-1}$ for x real
$y = u(x) = 2^{x-1} + 1$ for x real

And the same is true, too, for the curve of another function $y = h(x) = 2^{|x-1|} + 1$. That is, shifting by 1 the curve of $y = f(x) = 2^{|x|}$ in the direction of the x-axis, and then, shifting by 1 in the direction of the y-axis, we can readily get the curve of $y = h(x) = 2^{|x-1|} + 1$.

Fig. G

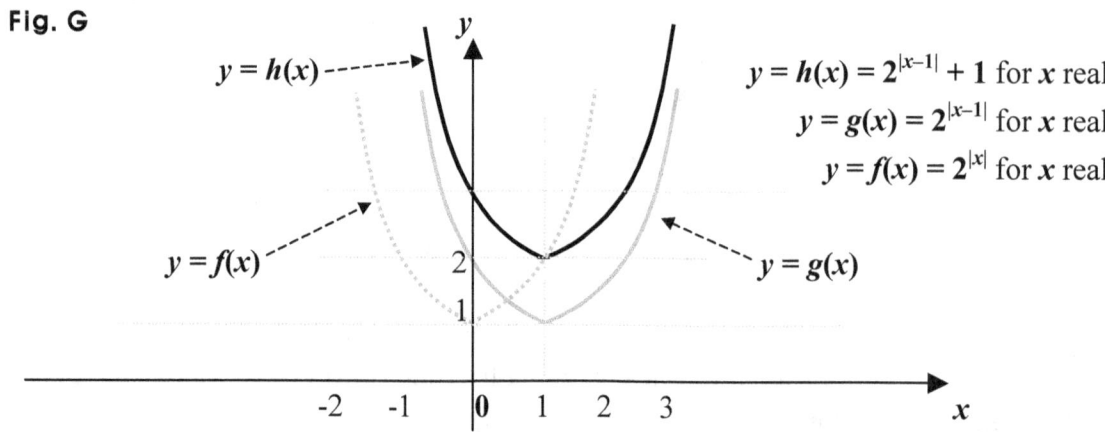

$y = h(x) = 2^{|x-1|} + 1$ for x real
$y = g(x) = 2^{|x-1|}$ for x real
$y = f(x) = 2^{|x|}$ for x real

How then, about the curve of this function: $y = h(x) = 2^{x-1} - 1$?

We can readily get it shifting by -1 the curve of a function $y = g(x) = 2^{x-1}$ in the direction of the y-axis.   How come?

Every time we subtract 1 from the output for a particular input of the function g, the difference is the output for the same particular input of the function h.

That is, we have: $h(x) = g(x) - 1$, because $h(x) = 2^{x-1} - 1$ and $g(x) = 2^{x-1}$. And thus, shifting by -1 the curve of g in the direction of the y-axis, we can get the curve of h.

Also, we know we can get the curve of $g(x) = 2^{x-1}$ shifting by 1 the curve of $y = f(x) = 2^x$ in the direction of the x-axis

So we can readily get the curve of $h(x) = 2^{x-1} - 1$ shifting by 1 the curve of $y = f(x) = 2^x$ in the direction of the x-axis, and then, shifting by -1 in the direction of the y-axis.

Fig. H

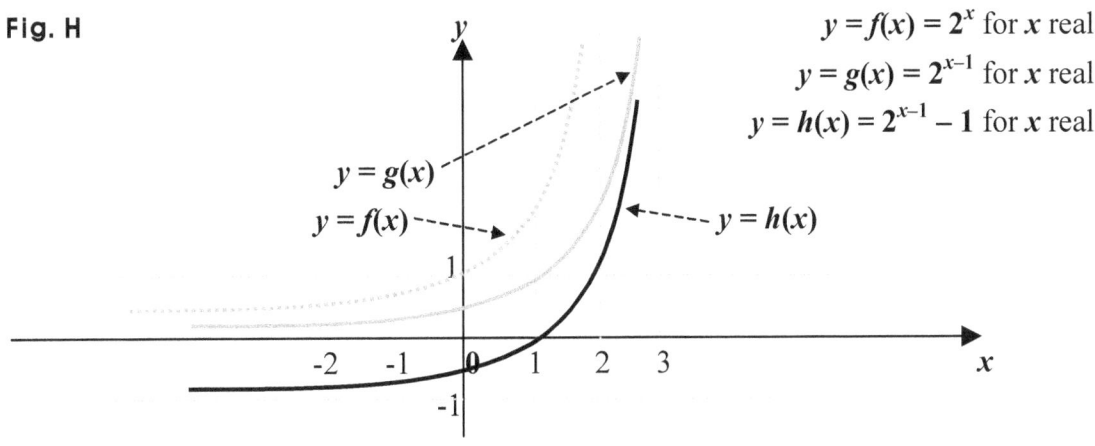

$y = f(x) = 2^x$ for x real
$y = g(x) = 2^{x-1}$ for x real
$y = h(x) = 2^{x-1} - 1$ for x real

And the same is true, too, for the curve of $y = h(x) = 2^{|x-1|} - 1$. That is, shifting by 1 the curve of $y = f(x) = 2^{|x|}$ in the direction of the x-axis, and then, shifting by -1 in the direction of the y-axis, we can readily get the curve of $y = h(x) = 2^{|x-1|} - 1$.

Fig. I

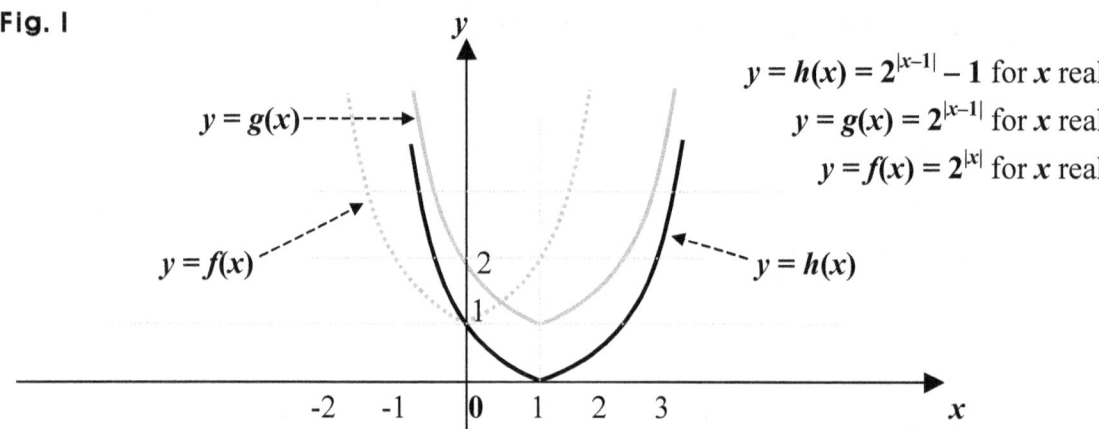

$y = h(x) = 2^{|x-1|} - 1$ for x real

$y = g(x) = 2^{|x-1|}$ for x real

$y = f(x) = 2^{|x|}$ for x real

$y = g(x)$

$y = f(x)$

$y = h(x)$

Next, how about the curve of another function $y = h(x) = 2^{|x+1|} - 2$?

We can readily get it shifting by -2 the curve of a function $y = g(x) = 2^{|x+1|}$ in the direction of the y-axis.   How come?

Every time we subtract 2 from an output for a particular input of the function g, the difference is the output for the same particular input of the function h.

That is, we have: $h(x) = g(x) - 2$, because $h(x) = 2^{|x+1|} - 2$ and $g(x) = 2^{|x+1|}$.

And thus, shifting by -2 the curve of g in the direction of the y-axis, we can get the curve of h.

And also, we can get the curve of $g(x) = 2^{|x+1|}$ shifting by -1 the curve of $y = f(x) = 2^{|x|}$ in the direction of the x-axis.

So shifting by -1 the curve of $y = f(x) = 2^{|x|}$ in the direction of the x-axis, and then, shifting by -2 in the direction of the y-axis, we can readily get the curve of:

$y = h(x) = 2^{|x+1|} - 2$.

Fig. J

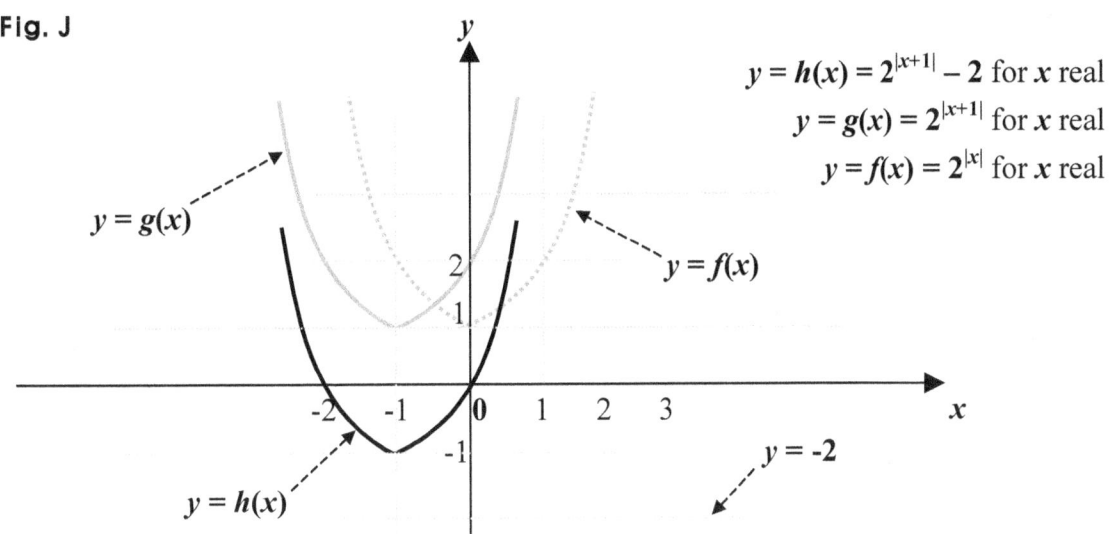

$$y = h(x) = 2^{|x+1|} - 2 \text{ for } x \text{ real}$$
$$y = g(x) = 2^{|x+1|} \text{ for } x \text{ real}$$
$$y = f(x) = 2^{|x|} \text{ for } x \text{ real}$$

$y = g(x)$

$y = f(x)$

$y = h(x)$

$y = -2$

Next, how about the curve of another function $y = h(x) = |2^{x+1} - 2|$?

We can readily get it shifting by -2 the curve of a function $y = 2^{x+1}$ in the direction of the y-axis, and then, move symmetrically about the x-axis, the portion of the curve below the x-axis.   How come?

We know: $|2^{x+1} - 2| \geq 0$ for any value of x.

So even though the values of $(2^{x+1} - 2)$ are negative for some values of x, the y-values are positive, that is, the y-coordinates are positive, because we have: $y = |2^{x+1} - 2|$.

And thus, no point in the curve of the function $y = h(x) = |2^{x+1} - 2|$ can be below the x-axis.

And in fact, those points that would be below the x-axis will be above the x-axis if there were no absolute sign.

So the curve has to be as shown below.

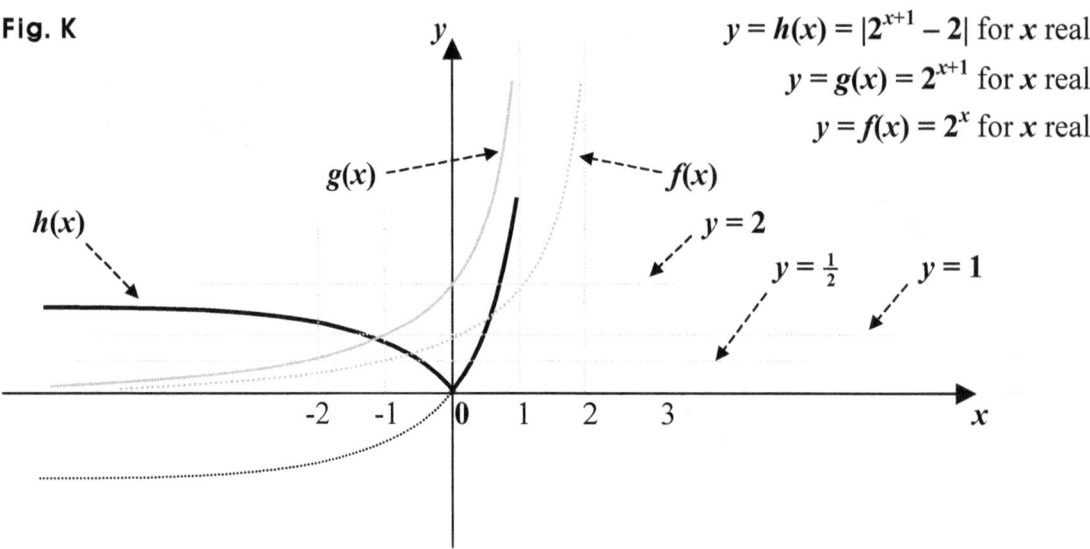

Fig. K

$y = h(x) = |2^{x+1} - 2|$ for x real

$y = g(x) = 2^{x+1}$ for x real

$y = f(x) = 2^x$ for x real

The curve solid in black is the curve of the function $y = h(x) = |2^{x+1} - 2|$ for x real.

How then, about the curve of another function $y = h(x) = |2^{|x+1|} - 2|$?

We can readily get it shifting by -2 the curve of a function $g(x) = 2^{|x+1|}$ in the direction of the y-axis, and then, move symmetrically about the x-axis, the portion of the curve below the x-axis.   How come?

We know: $|2^{|x+1|} - 2| \geq 0$.

So even though the values of $(2^{|x+1|} - 2)$ are negative for some values of x, the y-values are positive, that is the y-coordinates are positive, because $y = |2^{|x+1|} - 2|$.

So no point in the curve of the function $h(x) = |2^{|x+1|} - 2|$ can be below the x-axis.

And therefore, the curve has to be as shown below.

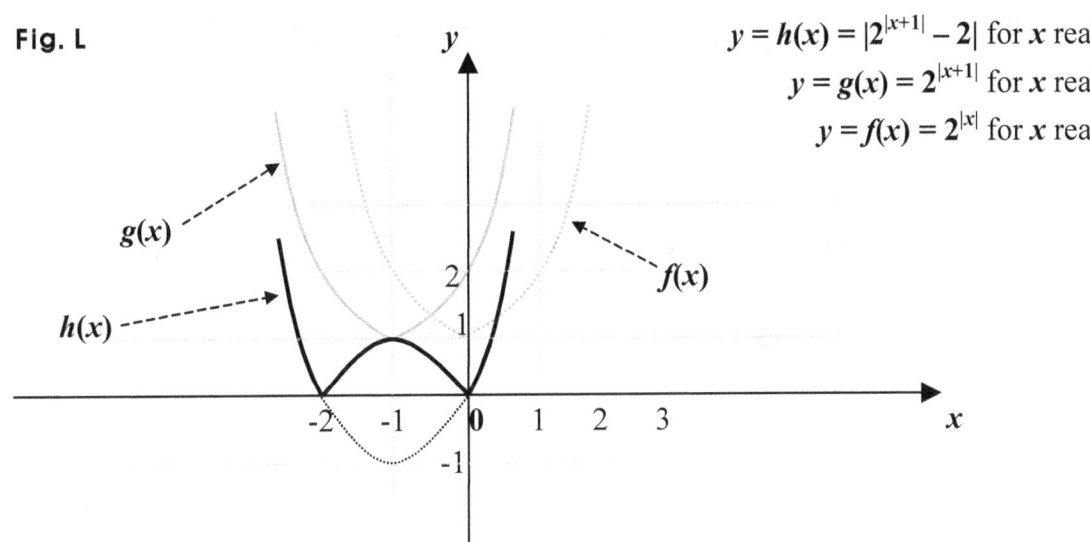

Fig. L

$y = h(x) = |2^{|x+1|} - 2|$ for x real

$y = g(x) = 2^{|x+1|}$ for x real

$y = f(x) = 2^{|x|}$ for x real

The curve solid in black is the curve of the function $h(x) = |2^{|x+1|} - 2|$.

So we can get the curve of a function with absolute sign the way below.

• Suppose we want to get the curve of a function $y = g(x) = |x|^2 - |x| - 2$.

Then first, get the curve of $y = p(x) = x^2 - x - 2$., and then, remove the portion on the left of the origin if such a portion exists, of course.

And next, copy and paste symmetrically about the y-axis, the portion on the right of the origin.

So getting the curve of $y = g(x) = |x|^2 - |x| - 2$, which is equal to $(|x| + 1)(|x| - 2)$, we can quickly get it the way below:

Fig. M

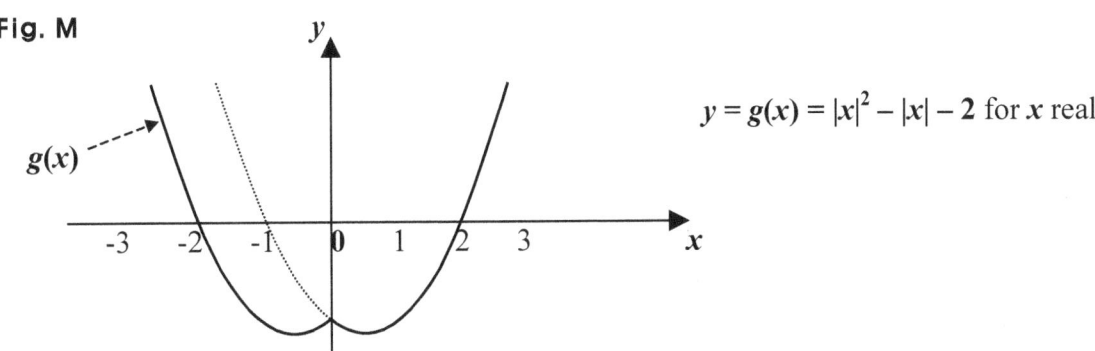

$y = g(x) = |x|^2 - |x| - 2$ for x real

• Suppose next, we want to get the curve of a function $y = |p(x)| = |x^2 - x - 2|$.

Then first, we get the curve of a function $y = p(x) = x^2 - x - 2$.

And next, if the curve has a portion below the *x*-axis, copy and paste the portion symmetrically about the *x*-axis, and then, remove the portion below the *x*-axis.

So we can get the curve of $y = |p(x)| = |x^2 - x - 2|$ the way below:

Fig. N

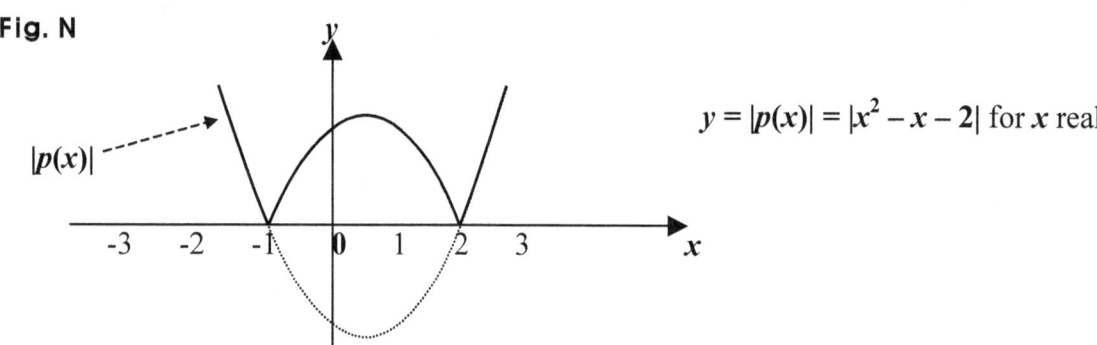

$y = |p(x)| = |x^2 - x - 2|$ for *x* real

If however, getting the curve of a function $y = h(x) = |x^2 - 2x + 2|$, we just get the curve of a function $y = q(x) = x^2 - 2x + 2$.

It's because no portion in the curve of $y = q(x)$ is below the *x*-axis. In fact, $q(x) = h(x)$. That is, $x^2 - 2x + 2 = |x^2 - 2x + 2|$, which doesn't need thus, absolute sign.

• Suppose next, we want to get the curve of $y = |p(|x|)| = ||x|^2 - 3|x| + 2|$.

Then, first, get the curve of $y = p(x) = x^2 - 3x + 2$, and then, remove the portion on the left of the origin if such a portion exists, of course.

Then again, in the curve of *p*, copy the portion on the right of the origin, and then, put on the left of the origin, the portion copied.

Then, we will get a curve symmetric about the *y*-axis, which is the curve of another function $y = p(|x|) = |x|^2 - 3|x| + 2$.
And next, if the curve of $p(|x|)$ has a portion below the *x*-axis, copy and paste the portion symmetrically about the *x*-axis, and then, remove the portion below the *x*-axis.

Then, we will get the curve we want, which is the curve of the function $|p(|x|)|$.

So for instance, putting in a graph the curve of $y = q(x) = ||x|^2 - |x| - 2|$, which equals $|(|x| + 1)(|x| - 2)|$, we can quickly get the curve of the function q the way below:

Beginning with the curve of $y = u(x) = x^2 - x - 2$, removing the portion on the left of the origin, and then, copying the portion on the right of the origin, and pasting the portion symmetrically about the y-axis, we get the curve of a function $y = v(x) = |x|^2 - |x| - 2$, which is shown below:

Fig. O

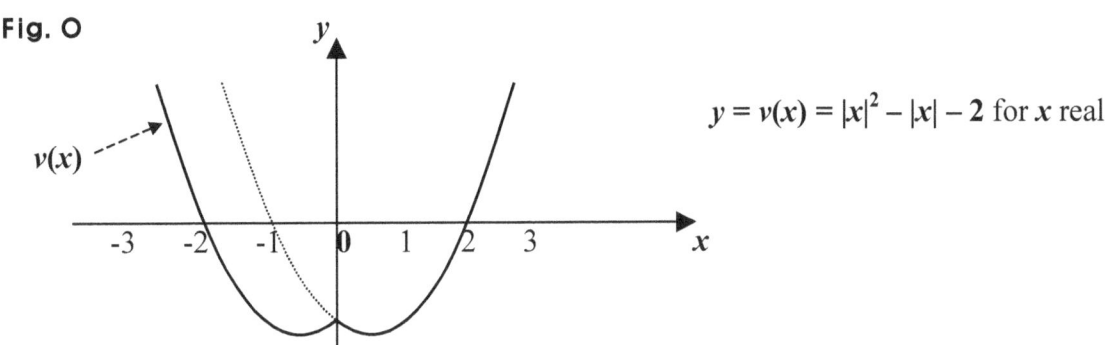

$y = v(x) = |x|^2 - |x| - 2$ for x real

And next, after copying, pasting, and removing the portion below the x-axis, we get the curve of the function $y = q(x) = ||x|^2 - |x| - 2|$, which is shown below:

Fig. P

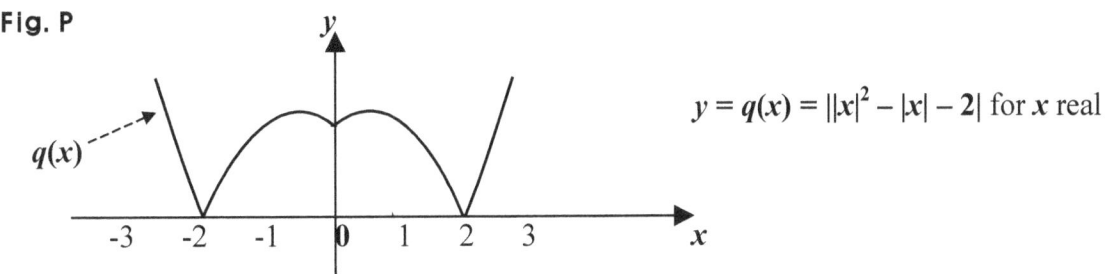

$y = q(x) = ||x|^2 - |x| - 2|$ for x real

If however, putting in a graph the curve of a function $y = s(x) = ||x|^2 - 2|x| + 2|$, we just get the graph of a function $y = t(x) = |x|^2 - 2|x| + 2$.   How come?

The curve of $y = t(x)$ does not have any portion below the x-axis. In fact, $t(x) = s(x)$. That is to say that we have: $|x|^2 - 2|x| + 2 = ||x|^2 - 2|x| + 2|$.

• How about then, the curve of a function $y = k(x) = |-|x|^2 + 2|x| - 2|$?

Beginning with a curve of a function $y = f(x) = -x^2 + 2x - 2$, removing the portion on the left of the origin, and then, copying the portion on the right of the origin, and pasting the portion symmetrically about the y-axis, we get the curve of $y = g(x) = -|x|^2 + 2|x| - 2$, and its curve will look like the one shown below:

Fig. Q

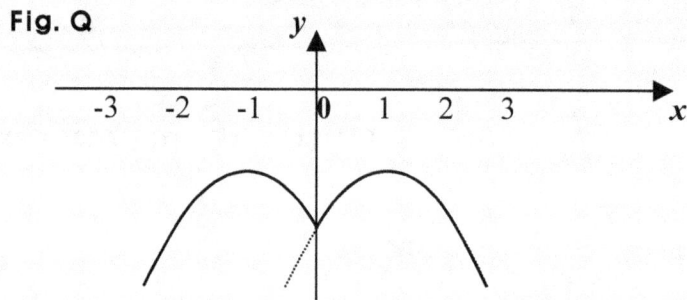

So next, since the entire curve above is below the x-axis, we just move the entire curve symmetrically about the x-axis. Then, we will get the curve of $y = k(x) = |-|x|^2 + 2|x| - 2|$, which will look like the one shown below:

Fig. R

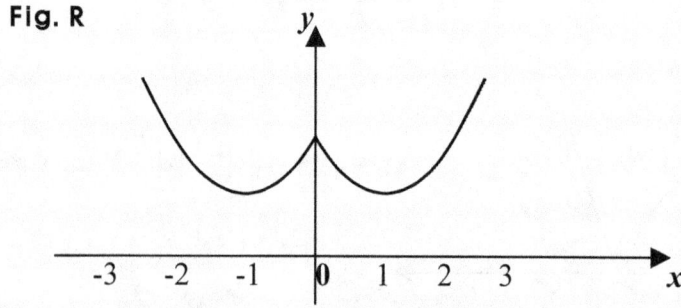

We can notice that the curve above is the same as the curve of $y = f(x) = ||x|^2 - 2|x| + 2|$, and is the same as the curve of $y = h(x) = |x|^2 - 2|x| + 2$, too.

That is to say that we have: $|-|x|^2 + 2|x| - 2| = ||x|^2 - 2|x| + 2| = |x|^2 - 2|x| + 2$.

So we have: $|-|x|^2 + 2|x| - 2| = |x|^2 - 2|x| + 2$.

• What then, about the curve of an equation $|y| = |x|^2 - 2|x| + 2$?

Putting in a graph an equation $y = x^2 - 2x + 2$, we get a parabola.

So let's begin with an equation $|y| = x^2 - 2x + 2$, the curve of which will be made of two parabolas symmetric about the x-axis.

Then, we want to consider two cases, one is a case where $y \geq 0$, and the other is: $y < 0$.

So first, if $y \geq 0$, we get: $|y| = x^2 - 2x + 2 \Rightarrow y = x^2 - 2x + 2 = (x - 1)^2 + 1$, and thus, the curve will be:

Fig. S

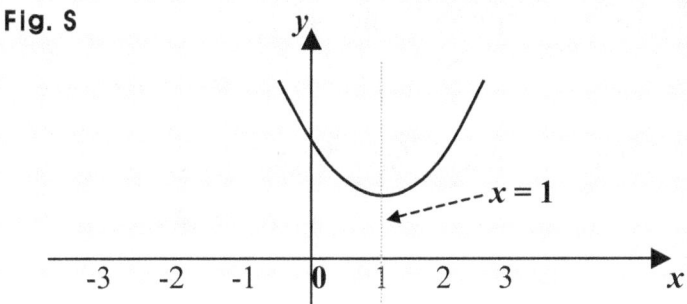

The line $x = 1$ is called the axis of symmetry, so the parabola above is symmetric about the line $x = 1$.

And next, if $y < 0$, we get: $|y| = x^2 - 2x + 2 \Rightarrow -y = x^2 - 2x + 2$

$\Rightarrow y = -x^2 + 2x - 2 = -(x - 1)^2 - 1$ for $y < 0$.

And thus, the curve will be as follows:

Fig. T

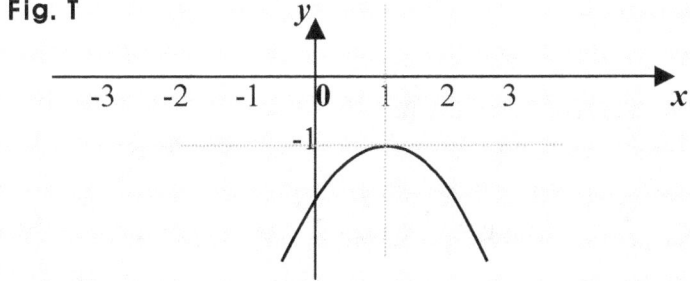

And thus, the curve of the equation $|y| = x^2 - 2x + 2$ is as follows:

Fig. U

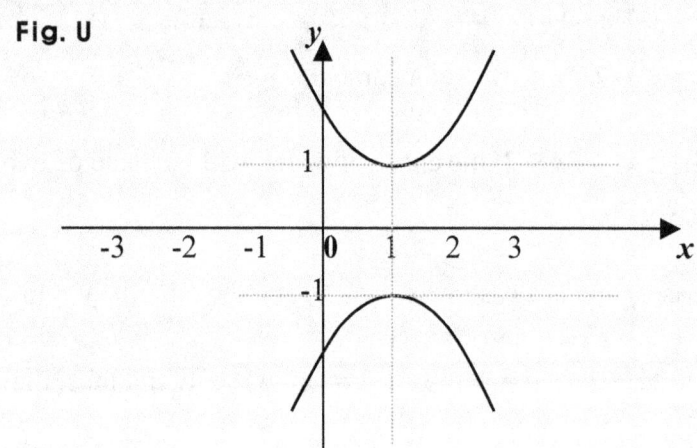

So the curve is made of two parabolas symmetric about the *x*-axis.

Let's next, move on to the curve of an equation $y = |x|^2 - 2|x| + 2$.

Then again, we want to consider two cases.

This time, one is a case where $x \geq 0$, and the other is: $x < 0$.

So first, if $x \geq 0$, we get: $y = |x|^2 - 2|x| + 2 \Rightarrow y = x^2 - 2x + 2 = (x - 1)^2 + 1$ for $x \geq 0$.

And thus, the curve will be as follows:

Fig. V

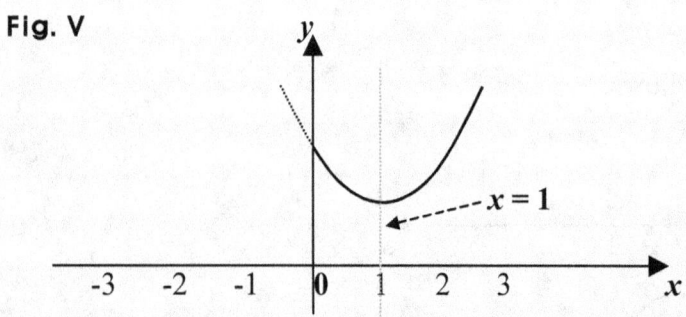

And next, if $x < 0$, we get: $y = |x|^2 - 2|x| + 2 \Rightarrow y = x^2 + 2x + 2 = (x + 1)^2 + 1$ for $x < 0$, and thus, the curve will be as below:

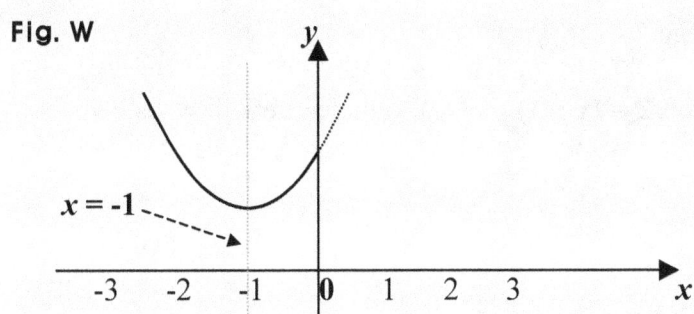

Fig. W

And thus, the curve of the equation $y = |x|^2 - 2|x| + 2$ is as follows:

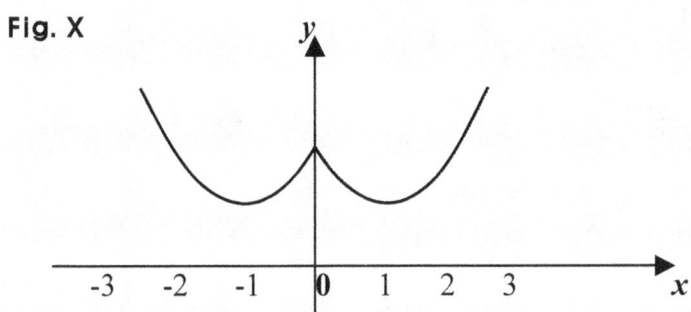

Fig. X

• Now, let's next, move on to the curve of the equation $|y| = |x|^2 - 2|x| + 2$.

Then, unlike the two equations above, we want to consider four cases.

One is a case where $y \geq 0$ and $x \geq 0$, another is a case where $y \geq 0$ and $x < 0$, another is a case where $y < 0$ and $x \geq 0$, and the other is a case where $y < 0$ and $x < 0$.

So first, if $y \geq 0$ and $x \geq 0$, we get: $|y| = |x|^2 - 2|x| + 2 \Rightarrow y = x^2 - 2x + 2 = (x - 1)^2 + 1$, and thus, the curve will be as follows:

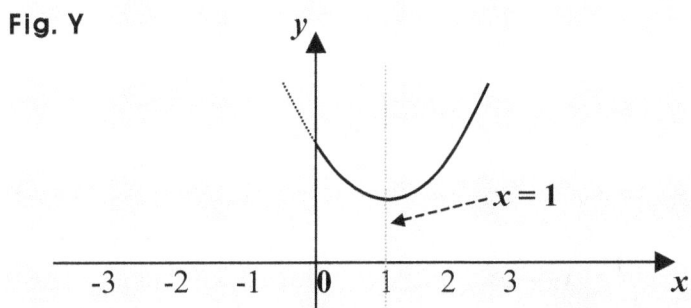

Fig. Y

And next, if $y \geq 0$ and $x < 0$, we get:

$|y| = |x|^2 - 2|x| + 2 \Rightarrow y = x^2 + 2x + 2 = (x + 1)^2 + 1.$ So the curve will be as below:

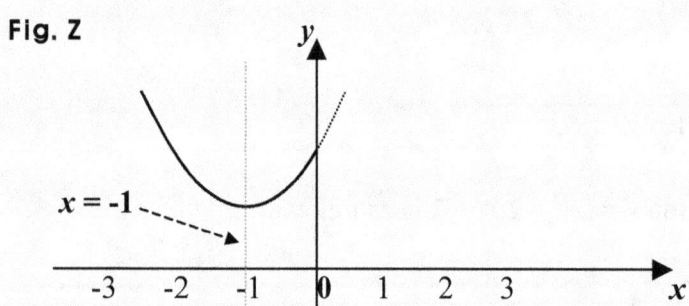

Fig. Z

And next, if $y < 0$ and $x \geq 0$, we get: $|y| = |x|^2 - 2|x| + 2 \Rightarrow -y = x^2 - 2x + 2$

$\Rightarrow y = -x^2 + 2x - 2 = -(x - 1)^2 - 1$, and thus, the curve will be as below:

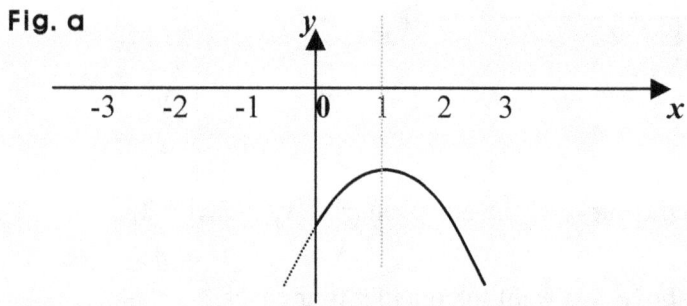

Fig. a

And next, if $y < 0$ and $x < 0$, we get: $|y| = |x|^2 - 2|x| + 2 \Rightarrow -y = x^2 + 2x + 2$

$\Rightarrow y = -x^2 - 2x - 2 = -(x + 1)^2 - 1$, and thus, the curve will be as below:

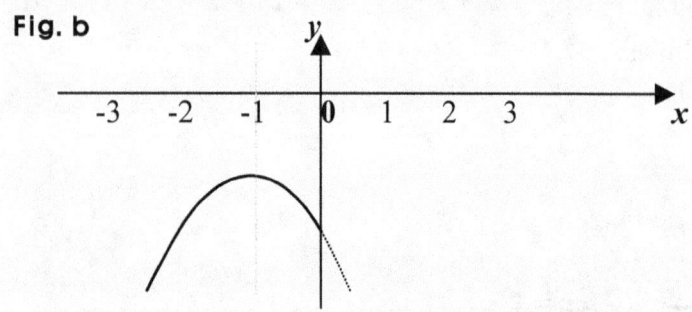

Fig. b

And thus, the curve of the equation $|y| = |x|^2 - 2|x| + 2$ is as follows:

Fig. c

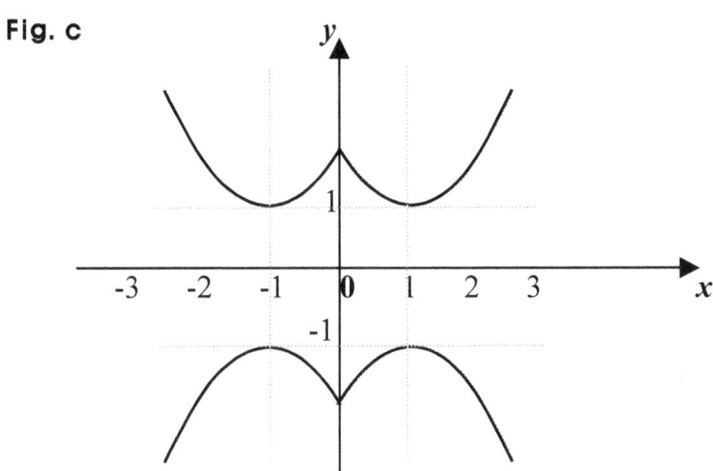

So the curve above is made of four curves.

And we can notice that if the curve in the first quadrant is *C*, the curve in the second quadrant is symmetric to *C* about the *y*-axis, the curve in the fourth quadrant is symmetric to *C* about the *x*-axis, and the curve in the third quadrant is symmetric to *C* about the origin.

And the same is true, too, for all the other equations of two variables.

So finding the curve of an equation $|y| = 2^{|x|}$, we want to consider the four cases as explained above.

Thus, first, if $y \geq 0$ and $x \geq 0$, we get: $|y| = 2^{|x|} \Rightarrow y = 2^x$, and thus, the curve will be:

Fig. d

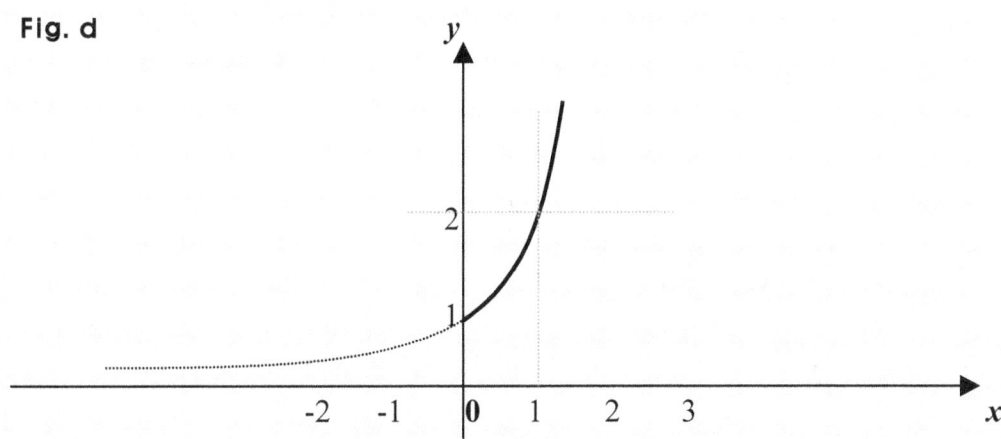

And next, if $y \geq 0$ and $x < 0$, we get: $|y| = 2^{|x|} \Rightarrow y = 2^{-x}$, and thus, the curve will be:

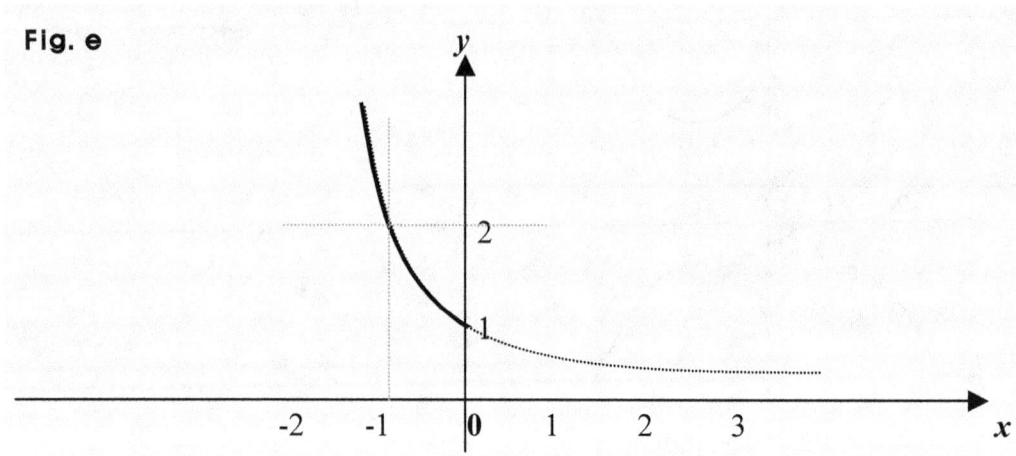

Fig. e

And next, if $y < 0$ and $x \geq 0$, we get:

$|y| = 2^{|x|} \Rightarrow -y = 2^x \Rightarrow y = -2^x$ for $y < 0$ and $x \geq 0$.

And thus, the curve will be:

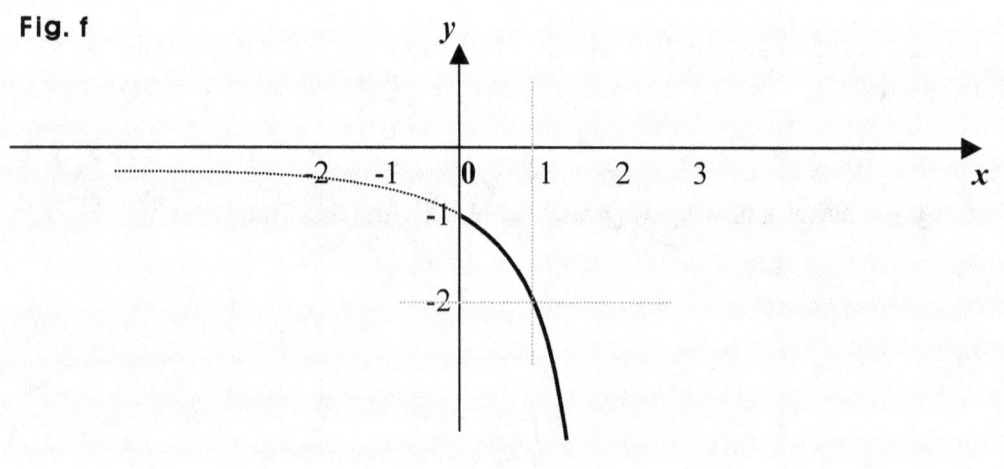

Fig. f

And next, if $y < 0$ and $x < 0$, we get:

$|y| = 2^{|x|} \Rightarrow -y = 2^{-x} \Rightarrow y = -2^{-x}$.   And thus, the curve will be as follows:

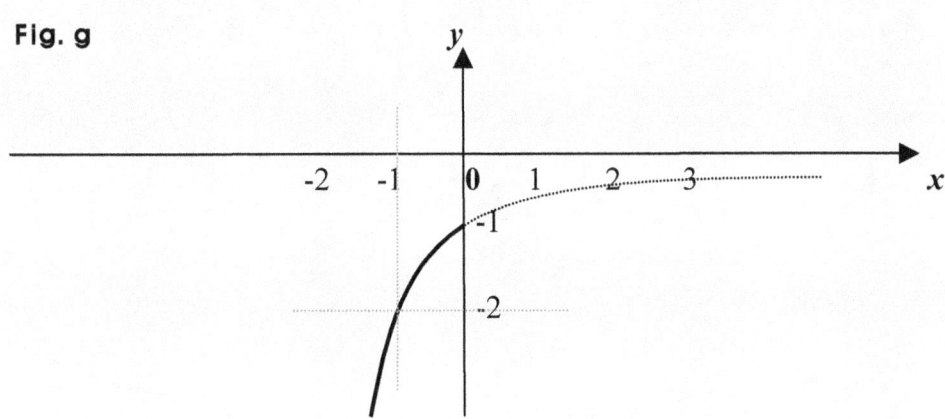

Fig. g

And thus, the curve of the equation $|y| = 2^{|x|}$ is as follows:

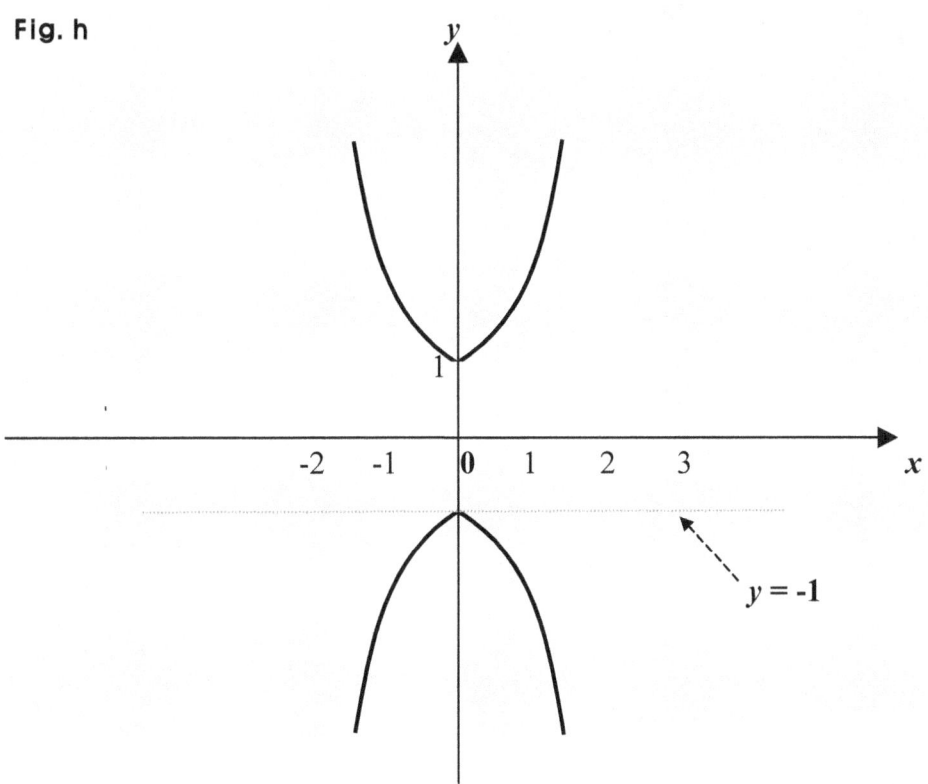

Fig. h

$y = -1$

So the curve above is made of four curves.

And if the curve in the first quadrant is **C**, the curve in the second quadrant is symmetric to **C** about the *y*-axis, the curve in the fourth quadrant is symmetric to **C** about the *x*-axis, and the curve in the third quadrant is symmetric to **C** about the origin.

Examples 2 in Exponential & Log Functions

Construct the graphs of the functions as follows:

0. $y = g(x) = \dfrac{2^x + 2^{-x}}{2}$ for x real.

1. $y = h(x) = \log_2 |x|$.

Suggestions or Solutions
To the **Problem** in the Example 0

Construct the graph of $y = g(x) = \dfrac{2^x + 2^{-x}}{2}$ for x real.

The given function $g(x)$ is an exponential function. What then, is the base?

We can put the function g this way: $y = g(x) = \frac{1}{2}(2^x + 2^{-x}) = 2^{x-1} + 2^{-x-1}$.

So assuming $g_1(x) = 2^{x-1}$, and $g_2(x) = 2^{-x-1}$, we can take g as the sum of the two functions g_1 and g_2.

And we know: $g_2(x) = 2^{-x-1} = 2^{-(x+1)} = (2^{-1})^{x+1}$.

So in the function g_1, the base is 2, and in the function g_2, the base is 2^{-1}.

How then, can we get the curve of the function g?

Setting $f_1(x) = 2^x$, and $f_2(x) = 2^{-x}$, we can notice that of the function $g(x)$, each output is the average of the two outputs from the two functions $f_1(x)$ and $f_2(x)$.

And thus, getting first, the two curves of f_1 and f_2, and next, taking the average of the two curves of f_1 and f_2, we can get the curve of g.   How then, do we get the average?

Taking the average of two outputs from f_1 and f_2, for the same input, we get the output of g for the same input. For instance, taking the average of $f_1(1)$ and $f_2(1)$, we get $g(1)$.

So from the two curves of f_1 and f_2, taking the midpoint between two points where the x-coordinates are the same, we can get a point in the curve of g.
And we do the same for all the other x-coordinates, too.

Then, connecting every midpoint of every pair of such two points in the curves of f_1 and f_2, we can get the average, which is the curve of the function g.

Now, let's to begin with, put in a graph the curve of $y = h(x) = a^x$, first. Then, we get:

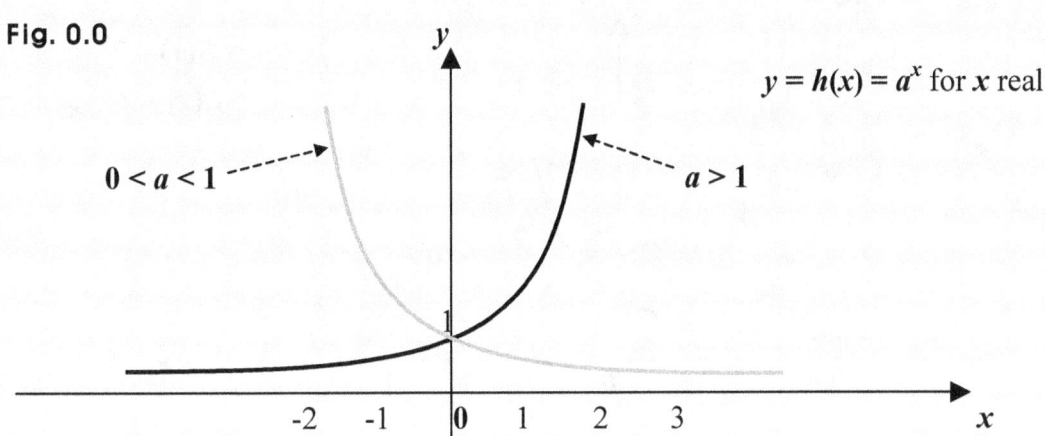

Fig. 0.0

$y = h(x) = a^x$ for x real

$0 < a < 1$

$a > 1$

So next, taking 2 as the base a, we can see that the curve below is the curve of $f_1(x) = 2^x$.

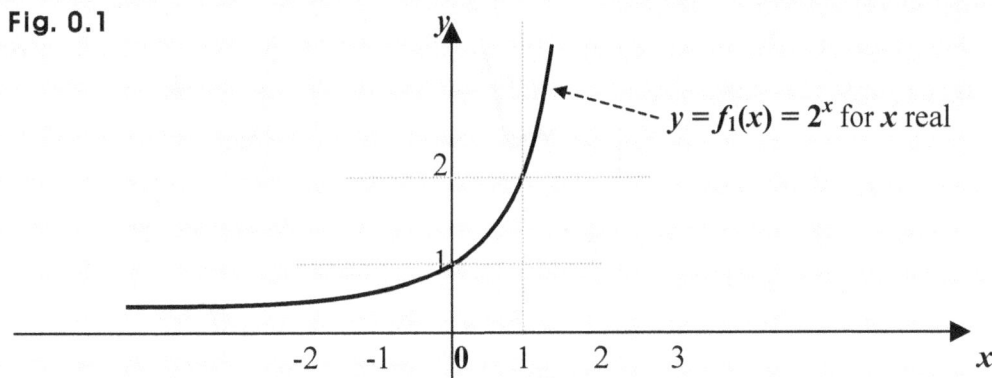

Fig. 0.1

$y = f_1(x) = 2^x$ for x real

And next, we have: $f_2(x) = 2^{-x} = (2^{-1})^x$. So in **Fig. 0.0**, taking 2^{-1} as the base a, we can see that the gray curve below is the curve of $f_2(x) = 2^{-x}$.

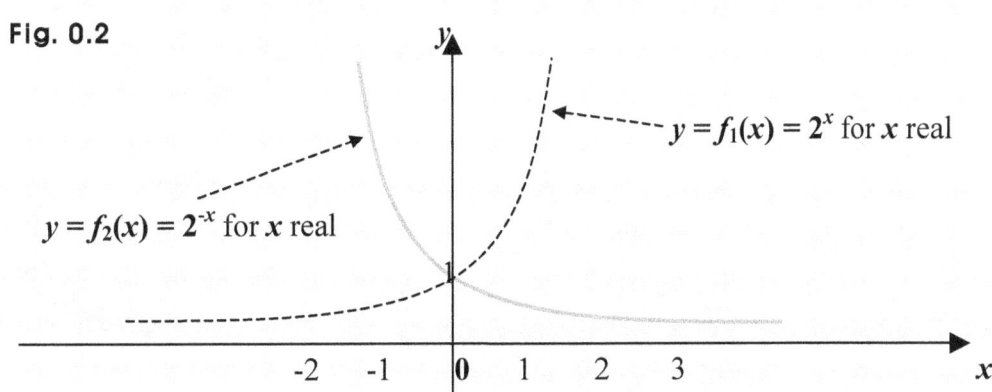

Fig. 0.2

$y = f_1(x) = 2^x$ for x real

$y = f_2(x) = 2^{-x}$ for x real

So next, putting in a graph the two curves of f_1 and f_2, we get:

Fig. 0.3

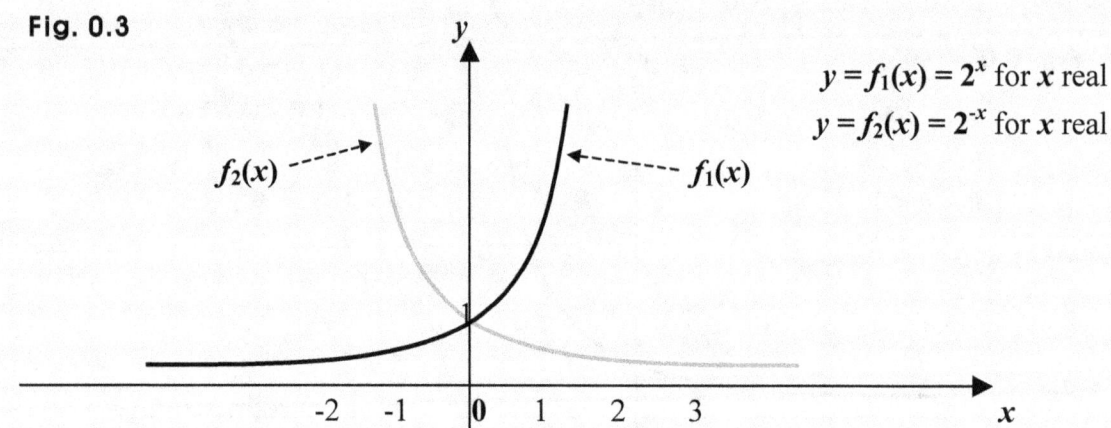

$y = f_1(x) = 2^x$ for x real
$y = f_2(x) = 2^{-x}$ for x real

So taking each of all the midpoints, we get the curve of $g(x)$, and the curve is as below:

Fig. 0.4

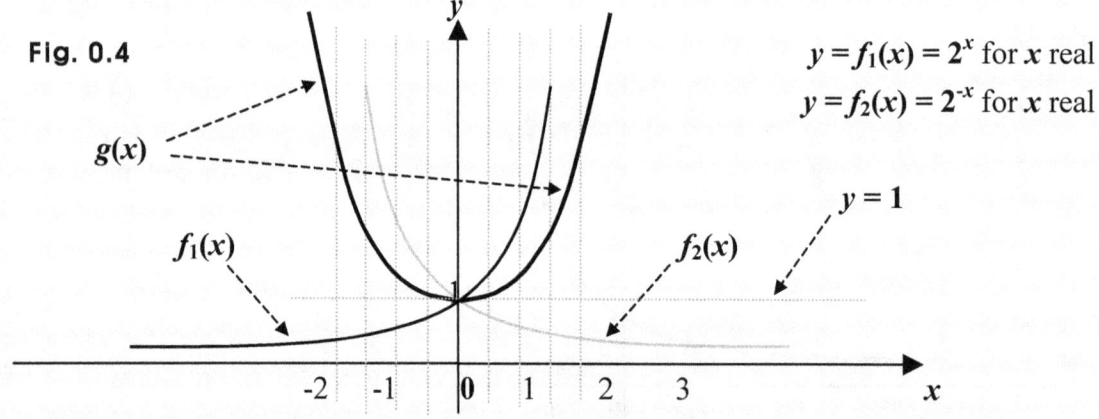

$y = f_1(x) = 2^x$ for x real
$y = f_2(x) = 2^{-x}$ for x real

$y = 1$

And if $s(x)$ is the sum of the two functions f_1 and f_2, the curve of $s(x)$ is as below:

Fig. 0.5

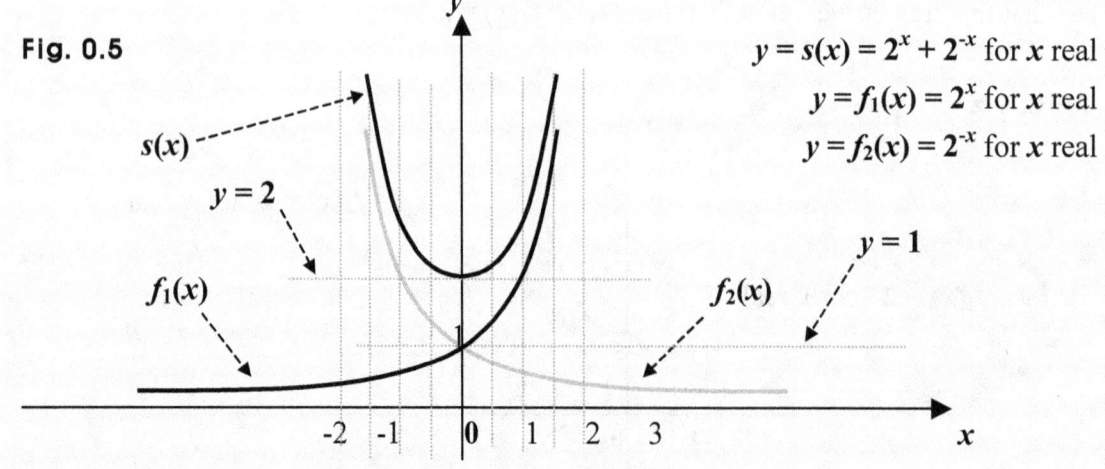

$y = s(x) = 2^x + 2^{-x}$ for x real
$y = f_1(x) = 2^x$ for x real
$y = f_2(x) = 2^{-x}$ for x real

$y = 1$

Suggestions or Solutions
To the **Problem** in the Example **1**

Construct the graph of $y = h(x) = \log_2 |x|$.

This time, we have a log function, where absolute sign is applied to the input variable. So getting the curve of this function: $y = f(x) = \log_2 x$ first, we can easily get the curve of the function h. To begin with, what is the domain of the function $f(x)$?

In a log, the antilog is positive, and in f, the antilog is x, so we get: $x > 0$, which is the domain. If not specified, the domain is the set of all the values the function can be defined. And the log function f can be defined for any x-value positive.
So the domain of f is: $x > 0$. And assuming R is a set of all real numbers, we can say that the domain of h is: $R - \{0\}$, because the antilog in a log cannot be 0.

So we can put the function h this way, too: $y = h(x) = \log_2 |x|$ for $x \neq 0$.

Let's now first, put in a graph, the curve of $y = u(x) = \log_a x$ for $x > 0$.   Then, we get:

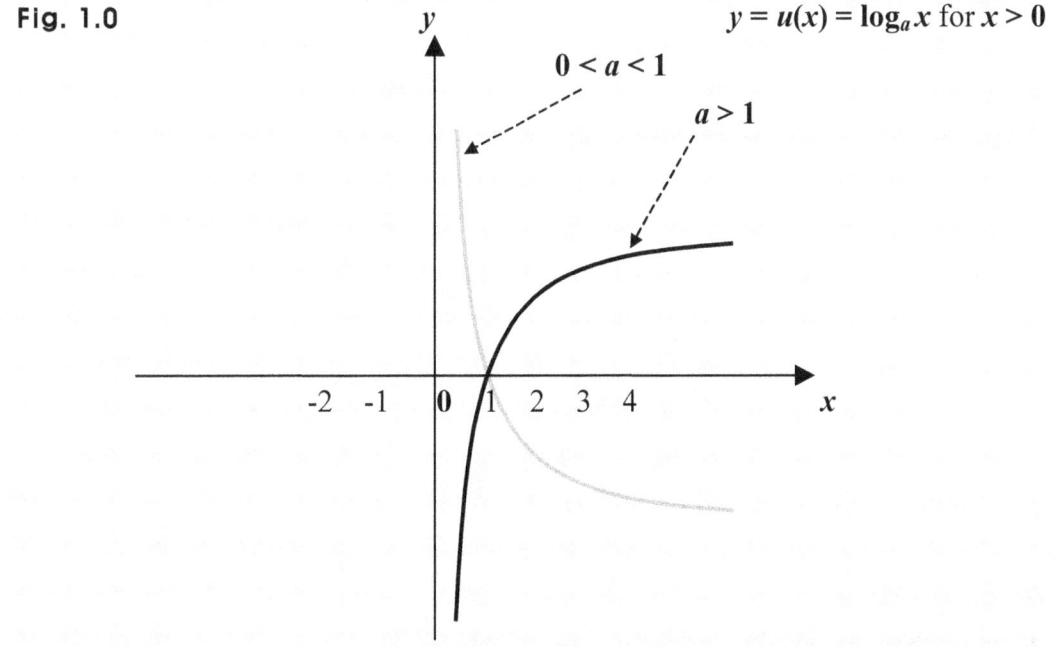

Fig. 1.0 y $y = u(x) = \log_a x$ for $x > 0$

$0 < a < 1$

$a > 1$

-2   -1   0   1   2   3   4 x

So taking 2 as the base *a*, we can see that the curve below is the curve of the function *f*.

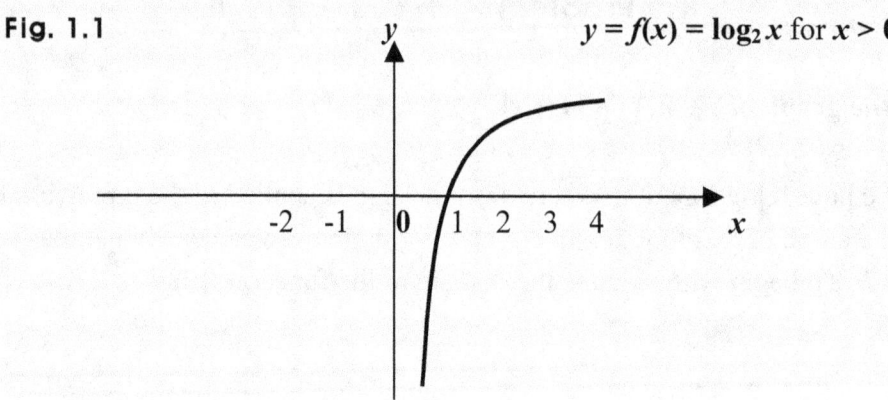

Fig. 1.1 $y = f(x) = \log_2 x$ for $x > 0$

So let's now move on to the function $y = h(x) = \log_2 |x|$.

In the **Examples 1**, we have covered the case below:

- Suppose we want to get the curve of a function $y = g(x) = |x|^2 - |x| - 2$.

Then first, get the curve of $y = p(x) = x^2 - x - 2$., and then, remove the portion on the left of the origin if such a portion exists, of course.

And next, copy and paste symmetrically about the *y*-axis, the portion on the right of the origin.

So we can get the curve of $y = h(x) = \log_2 |x|$ the way as follows:

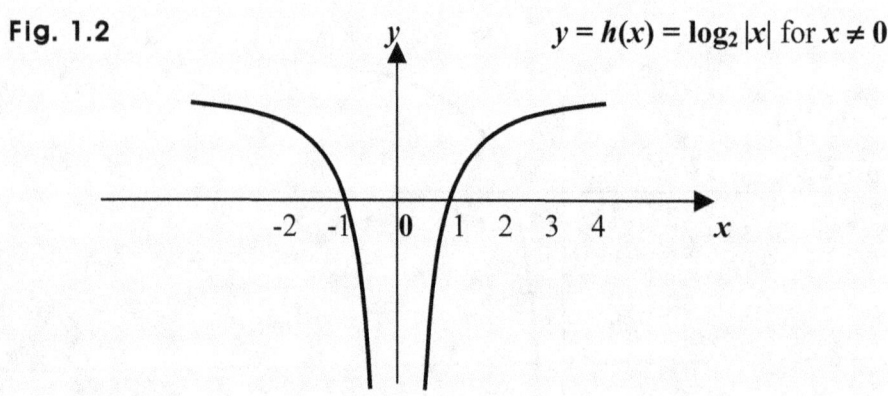

Fig. 1.2 $y = h(x) = \log_2 |x|$ for $x \neq 0$

- What then, about the curve of a function $y = k(x) = |\log_2 x|$?

In the **Examples 1**, we have covered the case below:

  • Suppose we want to get the curve of a function $y = |p(x)| = |x^2 - x - 2|$.

Then first, we get the curve of a function $y = p(x) = x^2 - x - 2$.

And next, if the curve has a portion below the x-axis, copy and paste the portion symmetrically about the x-axis, and then, remove the portion below the x-axis.

So we can get the curve of $y = k(x) = |\log_2 x|$ the way below:

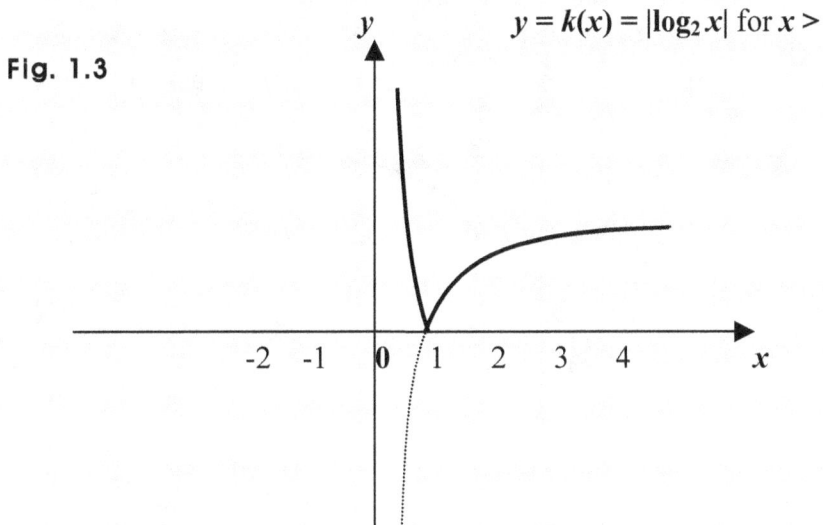

Fig. 1.3

$y = k(x) = |\log_2 x|$ for $x > 0$

  • And let's next, move on to a function $y = r(x) = |\log_2 |x||$.

In the set of **Examples 1**, we have covered the fact below:

  • Suppose we want to get the curve of $y = |p(|x|)| = ||x|^2 - 3|x| + 2|$.

Then, first, get the curve of $y = p(x) = x^2 - 3x + 2$, and then, remove the portion on the left of the origin if such a portion exists, of course.

Then again, in the curve of p, copy the portion on the right of the origin, and then, put on the left of the origin, the portion copied.

Then, we will get a curve symmetric about the y-axis, which is the curve of another function $y = p(|x|) = |x|^2 - 3|x| + 2$.

And next, if the curve of $p(|x|)$ has a portion below the x-axis, copy and paste the portion symmetrically about the x-axis, and then, remove the portion below the x-axis.

Then, we will get the curve we want, which is the curve of the function $|p(|x|)|$.

So we can quickly get the curve of $y = r(x) = |\log_2 |x||$, the way below:

Beginning with the curve of $y = m(x) = \log_2 x$, since no portion on the left of the origin, just copying the curve of m, and pasting it symmetrically about the y-axis, we get the curve of a function $y = n(x) = \log_2 |x|$, which is shown below:

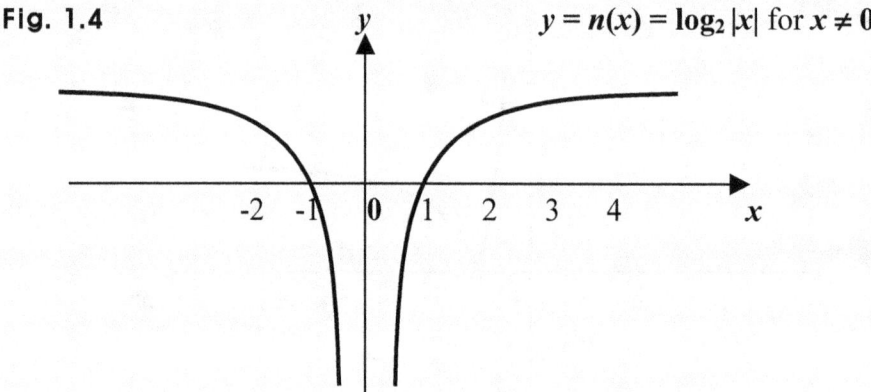

Fig. 1.4 $y = n(x) = \log_2 |x|$ for $x \neq 0$

And next, after copying, pasting, and removing the portion below the x-axis, we get the curve of the function $y = r(x) = |\log_2 |x||$, which is shown below:

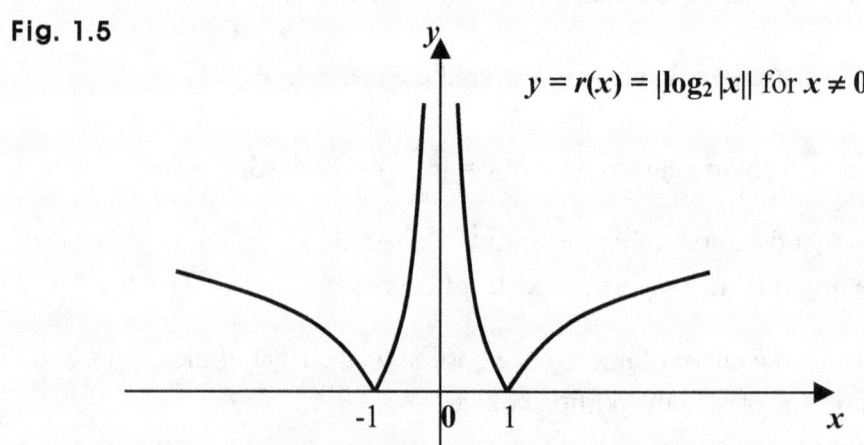

Fig. 1.5 $y = r(x) = |\log_2 |x||$ for $x \neq 0$

• How about the curve of $y = g(x) = \log_2 (x - 1)$?

We can readily get it shifting by 1 the curve of a function $y = f(x) = \log_2 x$ in the direction of the x-axis.   How come?

Suppose in g, the x-value changes from 2 to 4, for instance.

Then, the value of $(x - 1)$ changes from 1 to 3, of course. And we have: $y = f(x) = \log_2 x$.

So when x changes from 1 to 3, the value of $f(x)$ is the same as the value of $g(x)$ when x changes from 2 to 4. Thus, for instance, $f(1) = g(2)$, $f(2) = g(3)$, and $f(3) = g(4)$.

So when x changes from 2 to 4, the curve made by $g(x)$ is exactly the same as the curve made by $f(x)$ when x changes from 1 to 3, and behaves in the same manner.

And thus, shifting by 1 the entire curve of f in the direction of the x-axis, we can get the entire curve of g.

Fig. 1.6

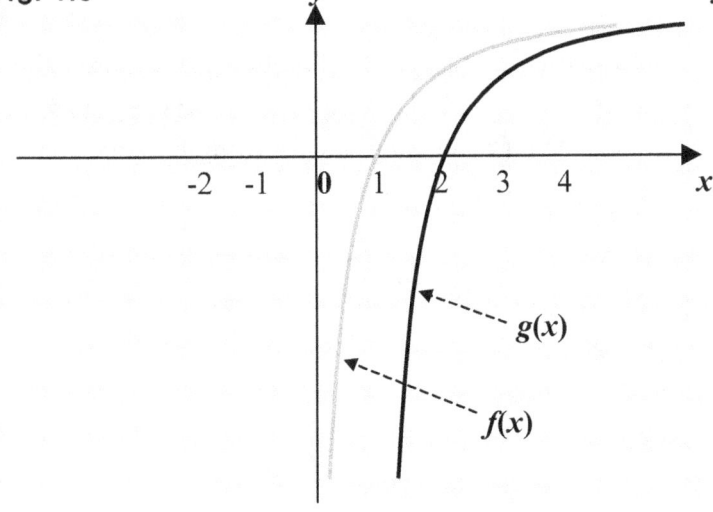

$y = g(x) = \log_2 (x - 1)$ for $x \neq 1$

$y = f(x) = \log_2 x$ for $x > 0$

• And the same is true, too, for the curve of a function $y = h(x) = \log_2 |x - 1|$.

That is to say that shifting by 1 the curve of a function $y = f(x) = \log_2 |x|$ in the direction of the x-axis, we can readily get the curve of $y = h(x) = \log_2 |x - 1|$.

So we can quickly get the curve of h the way below:

126

Fig. 1.7

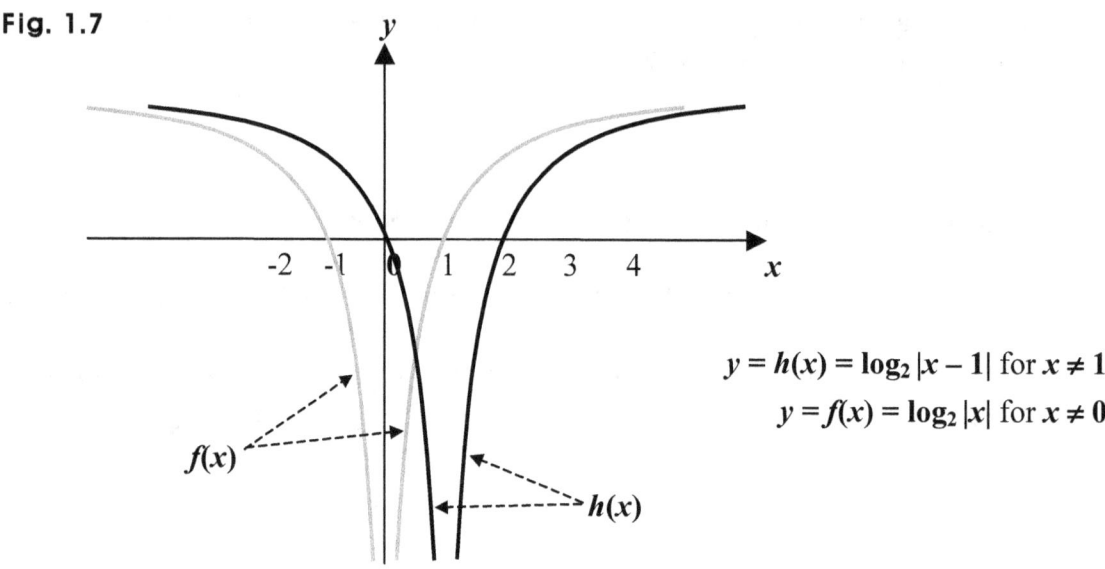

$$y = h(x) = \log_2 |x - 1| \text{ for } x \neq 1$$
$$y = f(x) = \log_2 |x| \text{ for } x \neq 0$$

$f(x)$

$h(x)$

• And we can readily get the curve of $y = h(x) = \log_2(x - 1) + 1$ shifting by 1 the curve of a function $y = g(x) = \log_2(x - 1)$ in the direction of the y-axis.   How come?

Every time we add 1 to an output for a particular input in the function g, the sum is the output for the same particular input in the function h. That is, we have: $h(x) = f(x) + 1$ because $h(x) = \log_2(x - 1) + 1$ and $g(x) = \log_2(x - 1)$. And thus, shifting by 1 the curve of g in the direction of the y-axis, we can get the curve of h.

Fig. 1.8

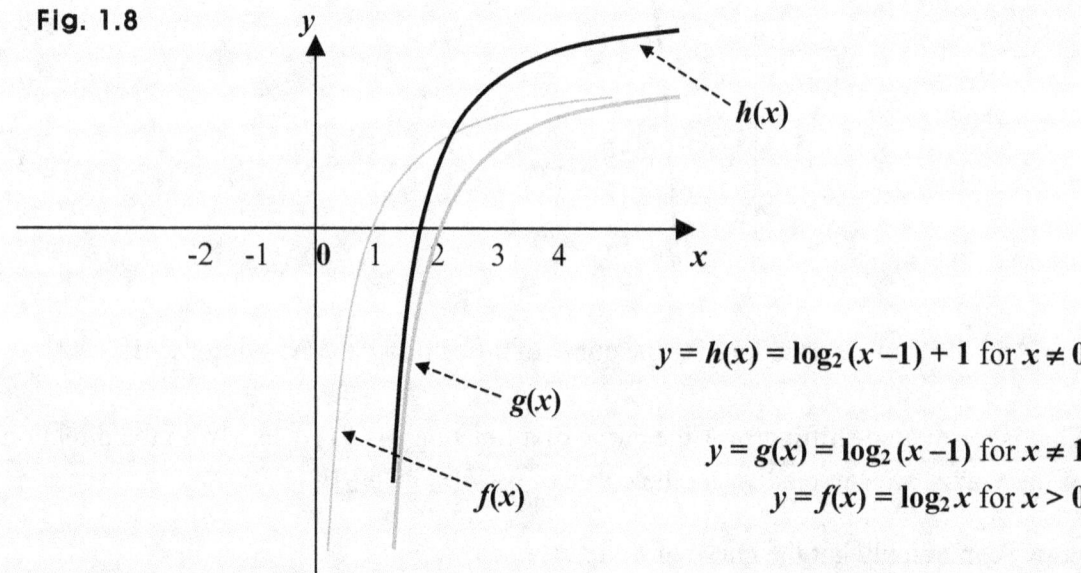

$h(x)$

$g(x)$

$f(x)$

$$y = h(x) = \log_2(x - 1) + 1 \text{ for } x \neq 0$$

$$y = g(x) = \log_2(x - 1) \text{ for } x \neq 1$$
$$y = f(x) = \log_2 x \text{ for } x > 0$$

• How about the curve of an equation $|y| = \log_2 |x|$?

Beginning with an equation $|y| = \log_2 x$, we want to consider two cases, one is a case where $y \geq 0$, and the other is: $y < 0$. So first, if $y \geq 0$, we get: $|y| = \log_2 x \Rightarrow y = \log_2 x$ for $y \geq 0$ and $x > 0$. and thus, the curve will be:

Fig. 1.9

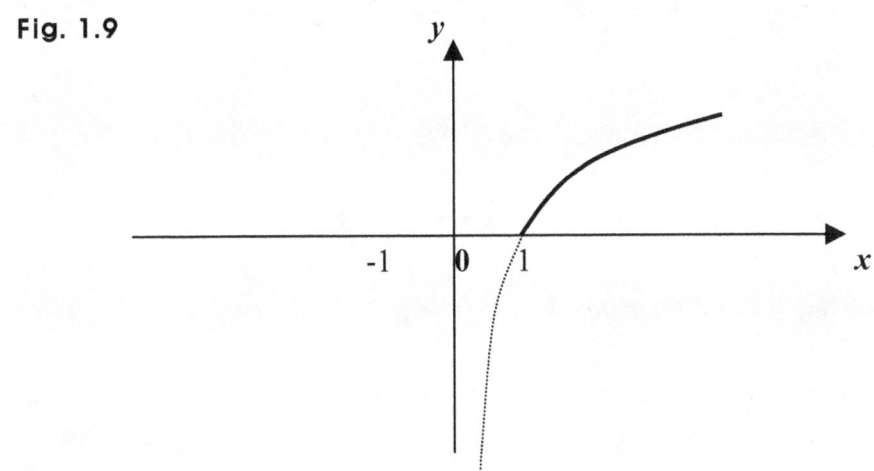

Next, if $y < 0$, we get: $|y| = \log_2 x \Rightarrow -y = \log_2 x \Rightarrow y = -\log_2 x$ for $x > 0$. So the curve is:

Fig. 1.A

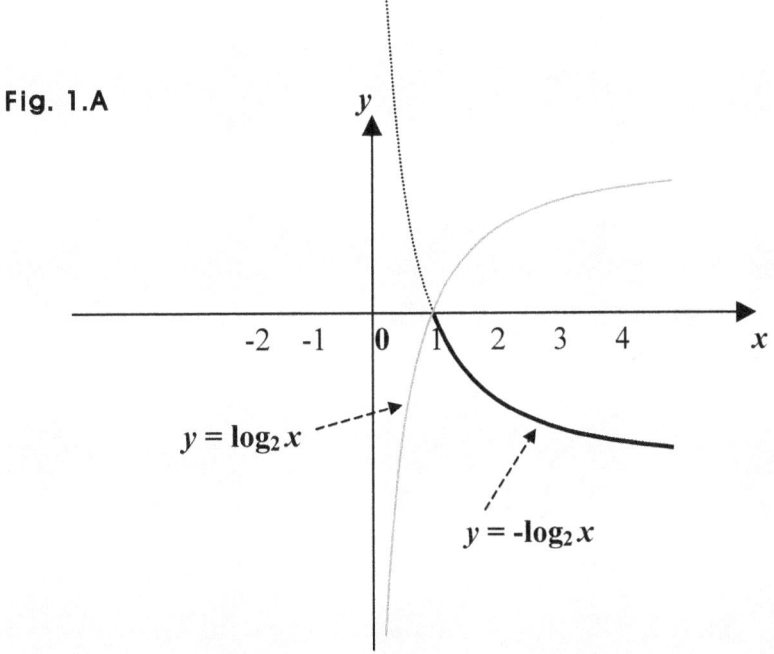

$y = \log_2 x$

$y = -\log_2 x$

And thus, the curve of the equation $|y| = \log_2 |x|$ is as follows:

Fig. 1.B

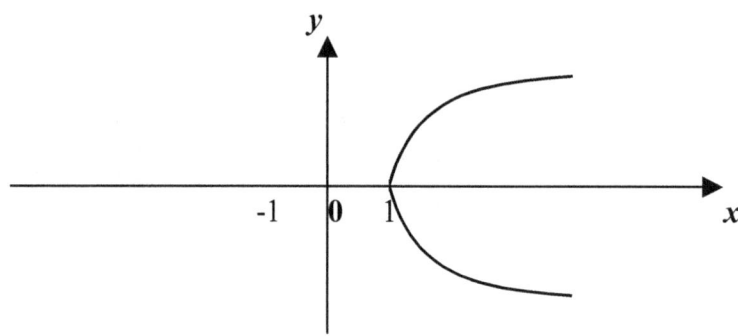

• Now, let's next, move on to the curve of the equation $|y| = \log_2 |x|$.

Then, unlike the two equations above, we want to consider four cases.

We know in a log, the antilog is positive.

So one is a case where $y \geq 0$ and $x > 0$, another is a case where $y \geq 0$ and $x < 0$, another is a case where $y < 0$ and $x > 0$, and the other is a case where $y < 0$ and $x < 0$.

So first, if $y \geq 0$ and $x \geq 0$, we get: $|y| = \log_2 |x| \Rightarrow y = \log_2 x$ for $y \geq 0$ and $x > 0$, and thus, the curve will be as follows:

Fig. 1.C

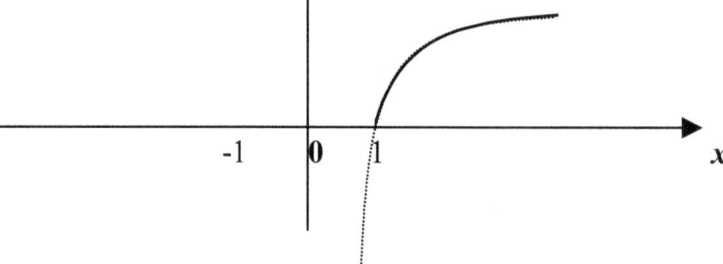

And next, if $y \geq 0$ and $x < 0$, we get: $|y| = \log_2 |x| \Rightarrow y = \log_2 (-x)$ for $y \geq 0$ and $x < 0$. So the curve will be as follows:

Fig. 1.D

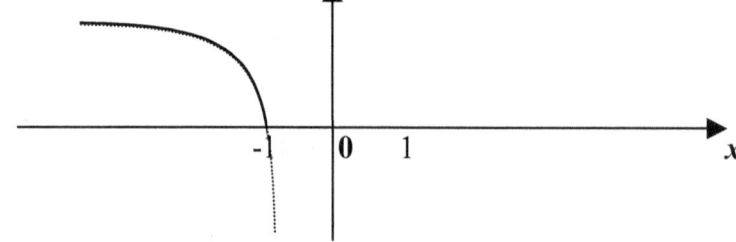

And next, if $y < 0$ and $x > 0$, we get:

$|y| = \log_2 |x| \Rightarrow -y = \log_2 x \Rightarrow y = -\log_2 x$ for $y < 0$, and $x > 0$. And thus, the curve will be:

Fig. 1.E

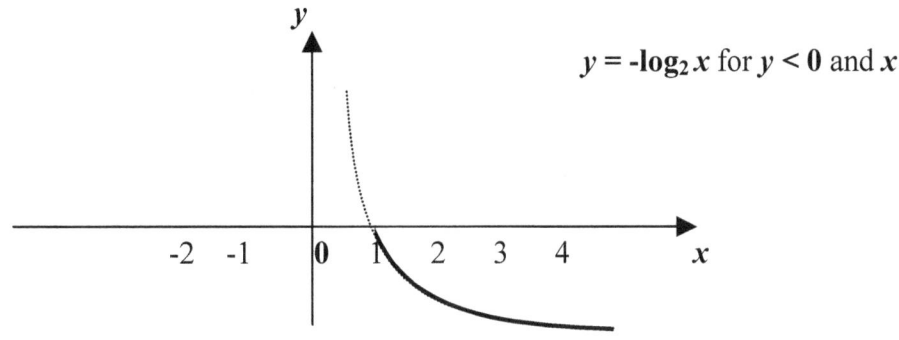

$y = -\log_2 x$ for $y < 0$ and $x > 0$

And next, if $y < 0$ and $x < 0$, we get:

$|y| = \log_2 |x| \Rightarrow -y = \log_2 (-x) \Rightarrow y = -\log_2 (-x)$ for $y < 0$ and $x < 0$. So the curve will be:

Fig. 1.F

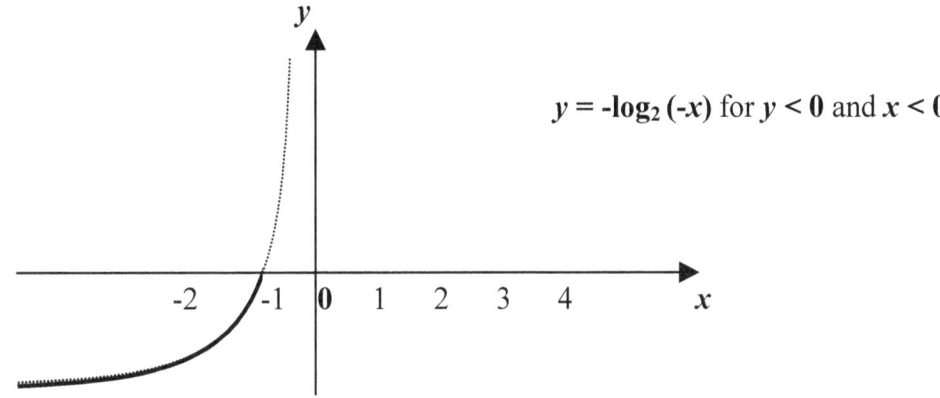

$y = -\log_2 (-x)$ for $y < 0$ and $x < 0$

And thus, the curve of $|y| = \log_2 |x|$ is as follows:

Fig. 1.G

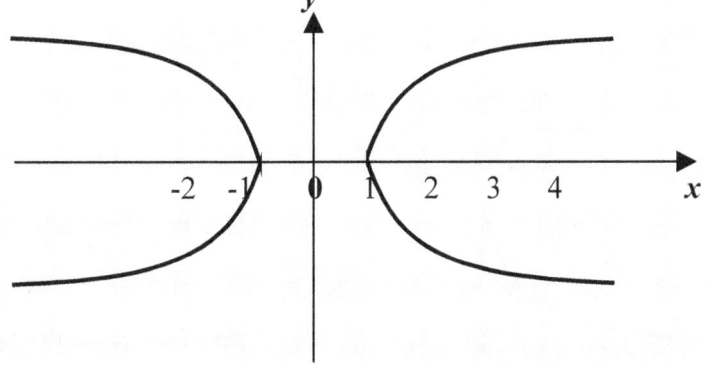

• What about the curve of another equation: $|y| = \log_2 |x + 1|$?

We can readily get it moving (translating) the entire curve above by -1 in the direction of the x-axis. If not quite sure, refer to **Examples 1**.

And by the similar fashion, getting the curve of an equation: $|y| = \log_2 |x - 1|$, we can readily get it moving (translating) the entire curve above by 1 in the direction of the x-axis. And the curve is shown below.

Fig. 1.H

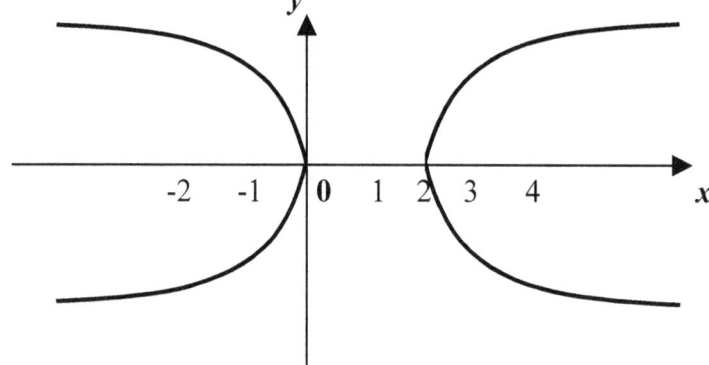

• What about the curve of another equation: $|y - 1| = \log_2 |x + 1|$?

We can readily get it moving (translating) the entire curve above by -1 in the direction of the y-axis, and the curve is shown below.

(If not sure, refer to **Examples 1** or the section on **Parallel Transformations** in the book, **GRAPH OPERATIONS**.)

Fig. 1.I

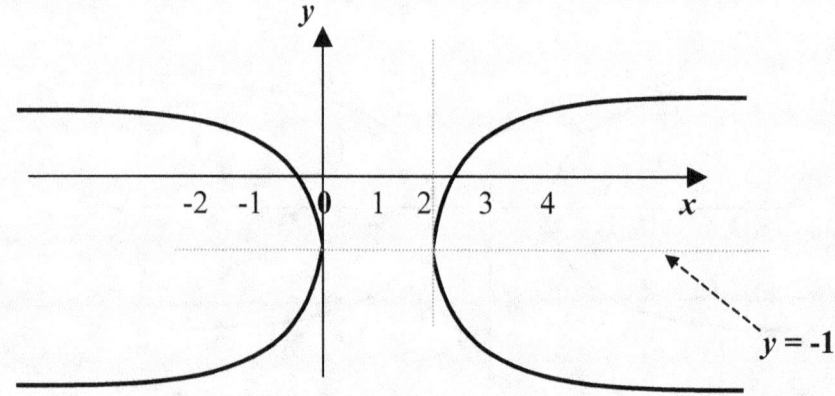

Examples 3 in Exponential & Log Functions

Though these examples are for your skill on functions, they are for your algebra skill, too. So following the processes in each example, pay attention to how expressions get changed at each step, and think about the idea behind the processes.

Once you've understood the idea and the processes, do each example yourself putting your solution processes in your writing. And the same is true, too, of curse, for all the other examples in this book.

Cliché but Always True: knowing is one thing, and Doing it is another.

Find the inverse of each of the functions as follows:

0. $f(x) = 6 \cdot 2^{x-1}$ for x real.

1. $g(x) = \dfrac{e^x - e^{-x}}{5}$ where e is a constant greater than 1.

2. $h(x) = \log_3 (3x + \sqrt{9x^2 - 1})$ for $x \geq 1/3$.

3. $k(x) = \log_3 (\frac{x}{3} + \sqrt{\frac{x^2-1}{9}}) + 1$ for $x \geq 1$.

Suggestions or Solutions
To the **Problem** in the Example **0**

Find the inverse of $f(x) = 6 \cdot 2^{x-1}$ for x real.

To begin with, $y = f(x) = 6 \cdot 2^{x-1} \Rightarrow \frac{y}{6} = 2^{x-1}$.

So next, by the definition for log functions, we can get: $\frac{y}{6} = 2^{x-1} \Leftrightarrow x - 1 = \log_2 \frac{y}{6}$.

Thus we get: $x - 1 = \log_2 \frac{y}{6} \Rightarrow x = \log_2 \frac{y}{6} + 1$.

So next, putting it in the x-y system, we get: $y = \log_2 \frac{x}{6} + 1$.

And next, we get: $6 \cdot 2^{x-1} > 0$ since x is real, and thus, the range of f is: $y > 0$.
So the domain of the log function is: $x > 0$.
And thus, assuming the log function is g, we get: $y = g(x) = \log_2 \frac{x}{6} + 1$ for $x > 0$.

And we can get the same the way below, too:

Simplifying the expression of the exponential function f, we can get:

$f(x) = 6 \cdot 2^{x-1} = 3 \cdot 2 \cdot 2^{x-1} = 3 \cdot 2^{x-1+1} = 3 \cdot 2^x$.   So we can set: $\frac{y}{3} = 2^x$.

And next, by the definition for log functions, we get: $\frac{y}{3} = 2^x \Leftrightarrow x = \log_2 \frac{y}{3}$.

So next, swapping the variables and assuming g is the log function, we get:

$y = g(x) = \log_2 \frac{x}{3}$ for $x > 0$.

If not quite sure of the idea behind the processes above, follow the steps below:

How do we find the inverse of a function?

To get straight to the point, finding the inverse, we put the input variable in terms of the output variable, and then swap the variables.

So finding the inverse of $y = f(x)$, we find an equation where x is set equal to an expression in terms of y, and then, swap x and y.

In short, finding the inverse of $y = f(x)$, we put x in terms of y, and then, swap x and y.

In other words, finding an equation where the input variable is set equal to an expression in terms of the output variable, and then, swapping the variables, we get the inverse.

How then, can we get such an equation?

To begin with, the definition for inverse functions is: $y = f(x) \Leftrightarrow x = f^{-1}(y)$.

The definition is saying that taking the inverse of f, we get: f^{-1}, read as f inverse. So f^{-1} is the inverse of f.   What then, do we mean by $f^{-1}(y)$?

It is <u>an expression in terms of y</u>. So for instance, it can be the case: $f^{-1}(y) = y^2 + 2y + 5$. And we have: $x = f^{-1}(y)$, which is the inverse of f.

(Note however, f^{-1} is just the name of the inverse, so we can use other letter as the name. For instance, we can use g as the name of the inverse. Then, we have: $x = g(y)$, the expression of which is the same as the expression of $f^{-1}(y)$. Thus, g is the inverse of f.)

So finding the inverse, what do we need to get?

We need to get the equation where x is equal to <u>an expression in terms of y</u>. And we can put the same idea the way below, too:

We know that in $y = f(x)$, x is the input variable, and y is the output variable. So if finding the inverse of f, we want to put the input variable in terms of the output variable. And putting the input variable in terms of the output variable, we get the equation where the input variable is set equal to an expression expressed in terms of the output variable.

How then in this problem, can we get such an equation stated above?

The function given in this problem is an exponential function, and taking the inverse of a function exponential, we get a log function.   So the inverse is a log function. How then, can we get the log function?

We can get it using the definition for logs, because the definition shows how functions of the two kinds are connected to each other. And the definition is as follows:

$y = b^x \Leftrightarrow x = \log_b y$, where $b > 0$, but $b \neq 1$.

How then, can we use the definition to get the log function, the solution to this problem?

Expressing the definition above in general, we can put it the way below:

- $p(y) = b^{q(x)} \Leftrightarrow q(x) = \log_b p(y)$, where $b > 0$, but $b \neq 1$.

And $p(y)$ is an expression in terms of y as $p(y) = 3y + 1$, and $q(x)$ is an expression in terms of x as $q(x) = 2x + 5$.

And in the case of the example above, using the definition for logs, we get:

$$3y + 1 = b^{(2x + 5)} \Leftrightarrow 2x + 5 = \log_b (3y + 1).$$

So we can get: $3y + 1 = b^{(2x + 5)} \Rightarrow 2x + 5 = \log_b (3y + 1)$, which is a log function.

Given therefore, an exponential function, we can get the inverse using the definition for logs, because the inverse is a log function.

And the same is true for the inverse of a log function, too. So given a log function, we can get the inverse using the definition for logs, because the inverse is an exponential function. And in that case, we can put the definition for logs the way below:

$$y = \log_b x \Leftrightarrow x = b^y, \text{ where } b > 0, \text{ but } b \neq 1.$$

And expressing the definition above in general, we can put it the way below:

$$p(y) = \log_b q(x) \Leftrightarrow q(x) = b^{p(y)}, \text{ where } b > 0, \text{ but } b \neq 1.$$

And of course, $p(y)$ is an expression in terms of y, and $q(x)$ is an expression in terms of x.

Now, in this problem, since we need to find the inverse of the exponential k, we want to find a log function using the function k, together with the definition for logs.

In short, finding the solution to this problem, we want to find a log function. How then, can we find the log function?

Finding the log function, we want to use the definition for log functions.
And then, we swap the variables, because we need to put the inverse in the *x-y* system, and not in the *y-x* system. That's simply because we usually put a function in the *x-y* system, where *x* is the input variable, and *y* is the output variable.

How do we use the definition though, in this problem, finding the log function?

Before using the definition, we want to put first, the exponential function f in the form stated above, and the form is: $p(y) = b^{q(x)}$, where b is constant, $p(y)$ is an expression in terms of y, and $q(x)$ is an expression in terms of x. So let's now put it in the form.

We have: $y = f(x) = 6 \cdot 2^{x-1}$. So putting it in the form of $p(y) = b^{q(x)}$, we get: $\dfrac{y}{6} = 2^{x-1}$.

It's because we can set: $p(y) = \dfrac{y}{6}$, and $q(x) = x - 1$.

So next, by the definition, we get: $\frac{y}{6} = 2^{x-1} \Leftrightarrow x - 1 = \log_2 \frac{y}{6}$.

Thus, we get: $x - 1 = \log_2 \frac{y}{6} \Rightarrow x = \log_2 \frac{y}{6} + 1$.

Next, swapping the variables, we get: $y = \log_2 \frac{x}{6} + 1$. We are not quite there yet, though.

We want to get the domain of the log function. How?

The domain is the range of the exponential function given, because the log function is the inverse of the exponential function.

We know that the domain of the exponential function f is a set of all real numbers.

And we get: $6 \cdot 2^{x-1} > 0$ if x can get all real numbers, and thus, the range of f is: $y > 0$.

So the domain of the inverse, that is, the domain of the log function is: $x > 0$.

And thus, assuming the log function is g, we get: $y = g(x) = \log_2 \frac{x}{6} + 1$ for $x > 0$.

And we can put it this way, too: $y = g(x) = \log_2 \frac{x}{3}$ for $x > 0$.

And we can have this, also: $6 \cdot 2^{x-1} = 3 \cdot 2^x$.
So we can notice that we can begin this problem with $f(x) = 3 \cdot 2^x$ instead of $f(x) = 6 \cdot 2^{x-1}$.

Then, we will get ended up with $y = g(x) = \log_2 \frac{x}{3}$ for $x > 0$.

How come though, it is the same as $y = g(x) = \log_2 \frac{x}{6} + 1$ for $x > 0$?

Simplifying: $\log_2 \frac{x}{6} + 1$, we get: $\log_2 \frac{x}{6} + 1 = \log_2 \frac{x}{6} + \log_2 2 = \log_2 (\frac{x}{6} \cdot 2) = \log_2 \frac{x}{3}$.

If not quite sure of the process above, refer to **POWERS AND LOGARITHMS**.

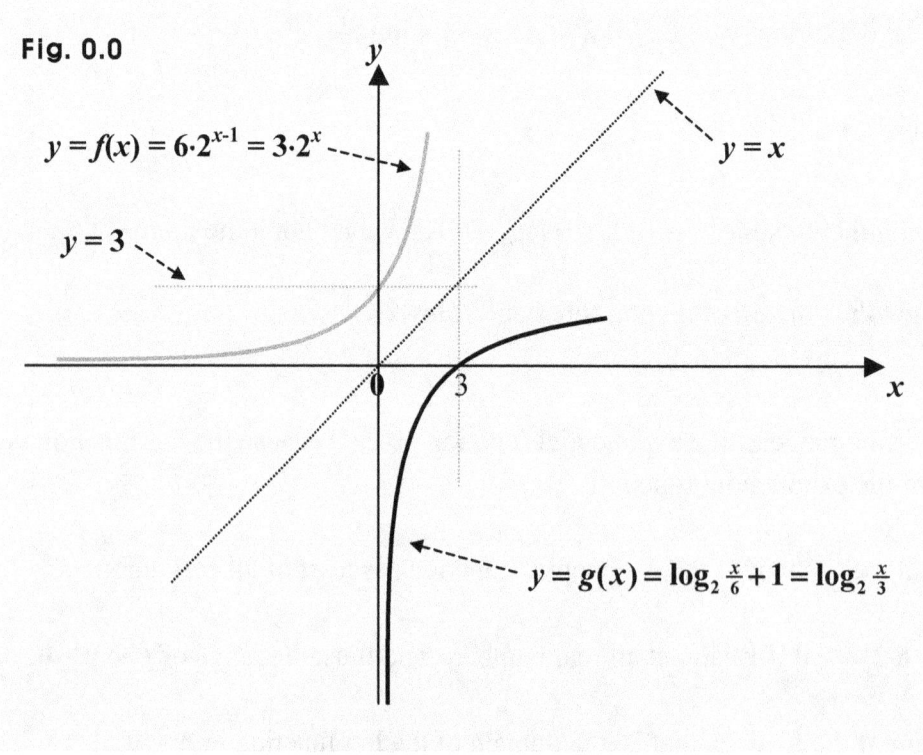

Fig. 0.0

The two curves are symmetric about the line $y = x$.

Suggestions or Solutions
To the **Problem** in the Example **1**

Find the inverse of $g(x) = \dfrac{e^x - e^{-x}}{5}$ **where e is a constant greater than 1.**

Setting: $t = e^x$, we get: $y = g(x) = \frac{t - t^{-1}}{5} = \frac{t^2 - 1}{5t} \Rightarrow 5ty = t^2 - 1 \Rightarrow t^2 - 5yt - 1 = 0$

$\Rightarrow t^2 - 5yt - 1 = t^2 - 5yt + (\frac{5y}{2})^2 - (\frac{5y}{2})^2 - 1 = (t - \frac{5y}{2})^2 - \frac{25y^2}{4} - 1 = (t - \frac{5y}{2})^2 - \frac{25y^2 + 4}{4} = 0$

$\Rightarrow (t - \frac{5y}{2})^2 = \frac{25y^2 + 4}{4} \Rightarrow t - \frac{5y}{2} = \frac{\pm\sqrt{25y^2 + 4}}{2} \Rightarrow t = \frac{\pm\sqrt{25y^2 + 4}}{2} + \frac{5y}{2} = \frac{5y \pm \sqrt{25y^2 + 4}}{2} \Rightarrow t = \frac{5y \pm \sqrt{25y^2 + 4}}{2}$.

We have: $5y \leq \sqrt{25y^2} < \sqrt{25y^2 + 4}$. So we get: $t = \frac{5y + \sqrt{25y^2 + 4}}{2}$ because $t > 0$.

Thus, we get: $e^x = \frac{5y + \sqrt{25y^2 + 4}}{2}$.

So next, by the definition for logs, we get: $x = \log_e \frac{5y + \sqrt{25y^2 + 4}}{2}$.

And thus, assuming h is the log function, and putting it in the x-y system, we get: $y = h(x) = \log_e \frac{5x + \sqrt{25x^2 + 4}}{2}$ for x real.

If not quite sure of the idea behind the processes above, follow the steps below:

Finding the inverse of g, we want to put x in terms of y, and of course, x and y are from the function g. Then, swapping the variables, we get the inverse.

In short, we want to get <u>an expression that is put in terms of the output variable y</u>.

How then, can we get such an expression?

To begin with, what function is g?

It is an exponential function.   So?

We know <u>the inverse of an exponential function is a log function</u>, which is therefore, the solution to this problem. And the definition for logs shows how functions of the two kinds are connected to each other. So we want to use the definition for logs.

And the definition is as follows: $y = b^x \Leftrightarrow x = \log_b y$, where $b > 0$, but $b \neq 1$.

And expressing the definition above in general, we can put it the way below:

- $p(y) = b^{q(x)} \Leftrightarrow q(x) = \log_b p(y)$, where $b > 0$, but $b \neq 1$.

And $p(y)$ is an expression in terms of y as $p(y) = 3y + 1$, and $q(x)$ is an expression in terms of x as $q(x) = 2x + 5$.

And in the case of the example above, using the definition for logs, we get:

$3y + 1 = b^{(2x + 5)} \Leftrightarrow 2x + 5 = \log_b (3y + 1)$.

So we can get: $3y + 1 = b^{(2x + 5)} \Rightarrow 2x + 5 = \log_b (3y + 1)$, which is a log function.

So in this problem, since we need to find the inverse of the exponential function g, we want to find a log function using the function g, together with the definition for logs.

In short, finding the solution to this problem, we want to find a log function.

And thus, we want to put first, the exponential function g in the form of $p(y) = b^{q(x)}$.

How then, do we put it in the form?

Putting it in the form is the key to this problem. And in fact, the same is true, too, for all the other problems in this kind.

Now, we have: $y = g(x) = \dfrac{e^x - e^{-x}}{5}$. The function g does not seem to get readily put in the form. How then, can we put it in the form?

We've got to do some algebra.
We can try first, taking as an unknown, the power e^x itself.
Then, we can get an equation for the unknown, and can try solving the equation.
Then, the solution will be an equality in a form of $p(y) = e^x$.
And then, we can try solving the equality for x, then swap the variables in the solution.

So to begin with, setting: $t = e^x$, we can set: $y = g(x) = \frac{t - t^{-1}}{5}$.

Then, we can put it this way: $y = \frac{1}{5}(t - \frac{1}{t}) = \frac{1}{5} \cdot \frac{t^2-1}{t} = \frac{t^2-1}{5t} \Rightarrow 5ty = t^2 - 1 \Rightarrow t^2 - 5yt - 1 = 0$,

which is quadratic, and thus, is not quite offensive. So next, solving it, we can get:

$$t^2 - 5yt - 1 = t^2 - 5yt + (\tfrac{5y}{2})^2 - (\tfrac{5y}{2})^2 - 1 = (t - \tfrac{5y}{2})^2 - \tfrac{25y^2}{4} - 1 = (t - \tfrac{5y}{2})^2 - \tfrac{25y^2+4}{4} = 0$$

$$\Rightarrow (t - \tfrac{5y}{2})^2 = \tfrac{25y^2+4}{4} \Rightarrow t - \tfrac{5y}{2} = \tfrac{\pm\sqrt{25y^2+4}}{2} \Rightarrow t = \tfrac{\pm\sqrt{25y^2+4}}{2} + \tfrac{5y}{2} = \tfrac{5y\pm\sqrt{25y^2+4}}{2} \Rightarrow t = \tfrac{5y\pm\sqrt{25y^2+4}}{2}.$$

And next, we know: $t = e^x$. So we now have: $e^x = \frac{5y\pm\sqrt{25y^2+4}}{2}$, which is therefore, in the form we want. What form? It is the form stated above: $p(y) = e^x$. What then, is the next?

We can not apply the definition for log functions to this: $e^x = \frac{5y\pm\sqrt{25y^2+4}}{2}$.
We've got two equations though.

One is: $e^x = \frac{5y+\sqrt{25y^2+4}}{2}$, and the other is: $e^x = \frac{5y-\sqrt{25y^2+4}}{2}$.

So we want to choose one from the two above. How?

Using the domain of the given function g, we can get the extent of e^x. Then, comparing the extent of e^x with the two extents of the two expressions: $\frac{5y+\sqrt{25y^2+4}}{2}$ and $\frac{5y-\sqrt{25y^2+4}}{2}$, we can make a choice. What then, is the domain of g?

We have: $y = g(x) = \dfrac{e^x - e^{-x}}{5}$, which can be defined for all x real.

So the domain is a set of all real numbers.

Thus, to begin with, for all x real, we get: $e^x > 0$. And we have: $t = e^x$. So we get: $t > 0$.

Thus next, getting first, the extents of the two expressions $\dfrac{5y \pm \sqrt{25y^2 + 4}}{2}$, we can begin with:

$5y \le \sqrt{25y^2} < \sqrt{25y^2 + 4}.$ Why not $5y = \sqrt{25y^2}$, though?

Going back to the function g above, we can see that the value of y can be negative, so $5y$ can be negative, but $\sqrt{25y^2} \ge 0$ no matter what value y may get.

So we get: $5y \le \sqrt{25y^2}$. And we know: $\sqrt{25y^2} < \sqrt{25y^2 + 4}$.

Thus next, for all values of y, we get: $\sqrt{25y^2 + 4} > 5y$.

And thus, we get: $5y + \sqrt{25y^2 + 4} > 0$, and $5y - \sqrt{25y^2 + 4} < 0$.

And we have: $t > 0$, and $t = \dfrac{5y \pm \sqrt{25y^2 + 4}}{2}$. So we get: $t = \dfrac{5y + \sqrt{25y^2 + 4}}{2}$.

And we have: $t = e^x$. Thus, we get: $e^x = \dfrac{5y + \sqrt{25y^2 + 4}}{2}$.

And the definition for logs is: $A = b^c \Leftrightarrow c = \log_b A$.

So next, by the definition for logs, we get: $x = \log_e \dfrac{5y + \sqrt{25y^2 + 4}}{2}$.

What then, is the next?

We want to get the domain, which is the range of the exponential function g.

We can quickly see the range of g using the graph of g. And the graph is as follows:

Fig. 1.0.

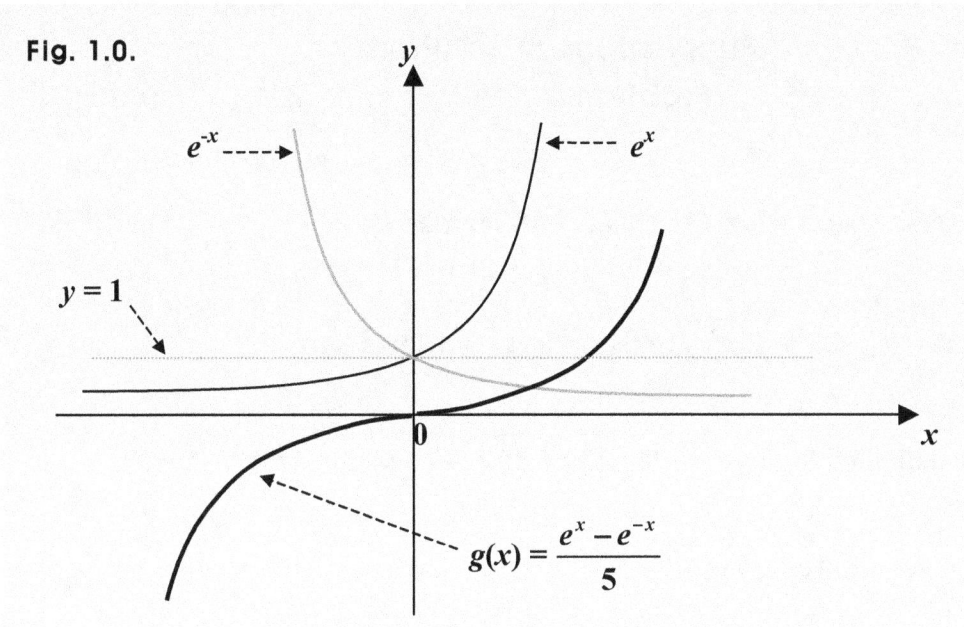

So we can see the range of **g** is a set of all real numbers, which is thus, the domain of the log function.   So assuming **h** is the log function, and swapping the variables, we get:

$$y = h(x) = \log_e \frac{5x + \sqrt{25x^2 + 4}}{2} \text{ for } x \text{ real.}$$

Fig. 1.1.

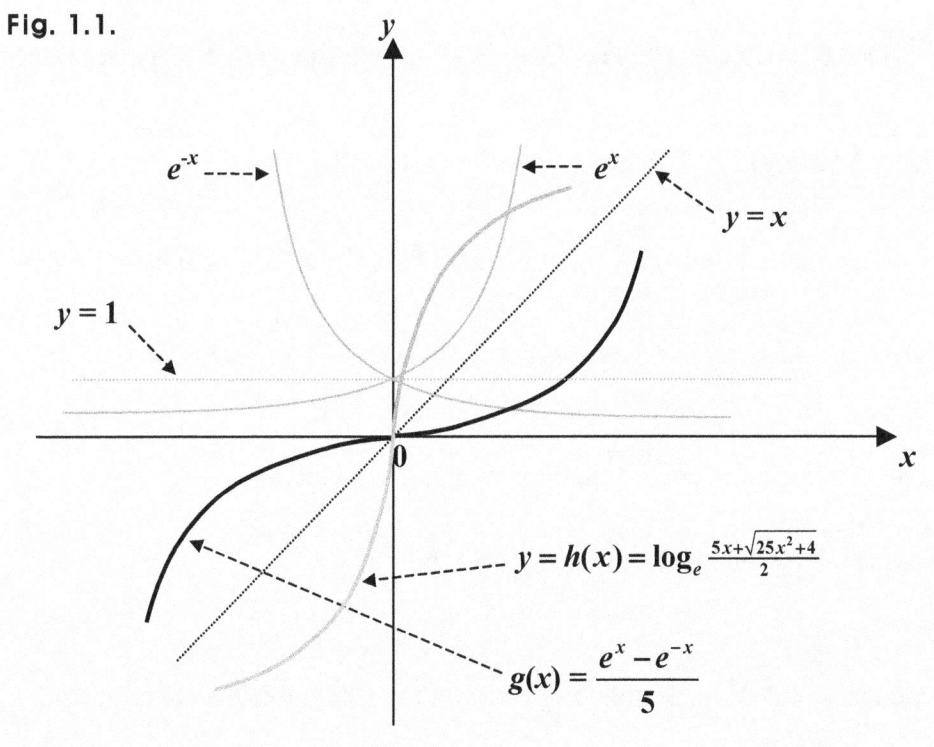

Suggestions or Solutions
To the **Problem** in the Example **2**

Find the inverse of $h(x) = \log_3(3x + \sqrt{9x^2 - 1})$ for $x \geq 1/3$.

First, $y = h(x) = \log_3(3x + \sqrt{9x^2 - 1}) \Rightarrow y = \log_3(3x + \sqrt{9x^2 - 1})$.

Next, by the definition for logs: $y = \log_3(3x + \sqrt{9x^2 - 1}) \Leftrightarrow 3x + \sqrt{9x^2 - 1} = 3^y$.

Setting $t = 3^y$, we get: $3x + \sqrt{9x^2 - 1} = t$.

$t - 3x = \sqrt{9x^2 - 1} \Rightarrow (t - 3x)^2 = 9x^2 - 1 \Rightarrow t^2 - 6tx + 9x^2 = 9x^2 - 1 \Rightarrow 6tx = t^2 + 1$
$\Rightarrow x = \frac{t}{6} + \frac{1}{6t}$.

So we get: $x = \frac{1}{6}(t + \frac{1}{t}) = \frac{1}{6}(3^y + \frac{1}{3^y}) = \frac{1}{6}(3^y + 3^{-y})$.

$x \geq 1/3 \Rightarrow 9x^2 - 1 \geq 0 \Rightarrow \sqrt{9x^2 - 1} \geq 0 \Rightarrow 3x + \sqrt{9x^2 - 1} \geq 1$, since $x \geq 1/3$. Thus, we get:

$\log_3(3x + \sqrt{9x^2 - 1}) \geq \log_3 1 = 0 \Rightarrow \log_3(3x + \sqrt{9x^2 - 1}) \geq 0$.

And thus, assuming the inverse is g, we get: $y = g(x) = \frac{1}{6}(3^x + 3^{-x})$ for $x \geq 0$.

If not quite sure of the idea behind the processes above, follow the steps below:

Taking the inverse of an exponential function, we get a log function.

So what do we get taking the inverse of a log function?

We get an exponential function. So this time, we want to find a function exponential.

Functions in both the kinds are inverse of each other, so we can find it by the definition for logs. And the definition is: $y = b^x \Leftrightarrow x = \log_b y$, where $b > 0$, but $b \neq 1$.

And we can put it this way, too: $y = \log_b x \Leftrightarrow x = b^y$, where $b > 0$, but $b \neq 1$.

And thus, getting the inverse either log or exponential, we can apply the definition above to the function we take the inverse of. Then, swapping the variables, we get the inverse.

The definition above is however, just a prototype if you will. So it's in the simplest form. And expressing the definition in general, we can put it the way as follows:

$p(y) = \log_b q(x) \Leftrightarrow q(x) = b^{p(y)}$, where $b > 0$, but $b \neq 1$.

And of course, $p(y)$ is an expression in terms of y as $p(y) = 3y + 1$, and $q(x)$ is an expression in terms of x as $q(x) = 2x + 5$.

And in the case of the example above, using the definition for logs, we get:

$3y + 1 = \log_b (2x + 5) \Leftrightarrow 2x + 5 = b^{(3y + 1)}$.

So we can get: $3y + 1 = \log_b (2x + 5) \Rightarrow 2x + 5 = b^{(3y + 1)}$, which is an exponential function.

So in this problem, since we need to find the inverse of the log function h, we want to find an exponential function using the function h, along with the definition for logs.

In short, finding the solution to this problem, we want to find a function exponential.

• Now in this problem, we have: $y = h(x) = \log_3 (3x + \sqrt{9x^2 - 1})$ for $x \geq 1/3$.

So the exponential function g is already in the form of $p(y) = \log_b q(x)$.

In this case, $p(y) = y$, and $q(x) = 3x + \sqrt{9x^2 - 1}$.

So next, by the definition, we can set: $y = \log_3 (3x + \sqrt{9x^2 - 1}) \Leftrightarrow 3x + \sqrt{9x^2 - 1} = 3^y$.

Thus next, swapping the variables, and naming the inverse, do we get the inverse?

It's not quite the case yet.

That's because just swapping the variables now, we just get: $3y + \sqrt{9y^2 - 1} = 3^x$, which is not the form we want, because the output variable y in the inverse is yet to be isolated.

That is, assuming the inverse is g, using the equality as is, we cannot put the inverse in a form of this: $y = g(x)$.   So we cannot just use this equality: $3x + \sqrt{9x^2 - 1} = 3^y$.

That is, we want to isolate x first.   How then, can we do that?

Assuming $t = 3^y$, we get: $3x + \sqrt{9x^2 - 1} = t$.   Then, we get:

$t - 3x = \sqrt{9x^2 - 1} \Rightarrow (t - 3x)^2 = 9x^2 - 1 \Rightarrow t^2 - 6tx + 9x^2 = 9x^2 - 1$

$\Rightarrow 6tx = t^2 + 1 \Rightarrow x = \frac{t}{6} + \frac{1}{6t}$.

And we know: $t = 3^y$.   So we get: $x = \frac{1}{6}(t + \frac{1}{t}) = \frac{1}{6}(3^y + \frac{1}{3^y}) = \frac{1}{6}(3^y + 3^{-y})$.

Thus, we get: $x = \frac{1}{6}(3^y + 3^{-y})$.   We are not quite there yet though.
What then, is the next?

We want to get the domain of the inverse, which is the range of the original function h. And the domain of h is: $x \geq 1/3$. So getting the range of the function h, we can get first:

$x \geq 1/3 \Rightarrow 9x^2 - 1 \geq 0 \Rightarrow \sqrt{9x^2 - 1} \geq 0 \Rightarrow 3x + \sqrt{9x^2 - 1} \geq 1$, since $x \geq 1/3$.

Thus, we get: $\log_3 (3x + \sqrt{9x^2 - 1}) \geq \log_3 1 = 0 \Rightarrow \log_3 (3x + \sqrt{9x^2 - 1}) \geq 0$.
So we can see that the range of h is: $y \geq 0$. And thus, the domain of the inverse is: $x \geq 0$.

So assuming the inverse is g and putting it in the x-y system, so swapping the variables, we get: $y = g(x) = \frac{1}{6}(3^x + 3^{-x})$ for $x \geq 0$.

Fig. 2.0

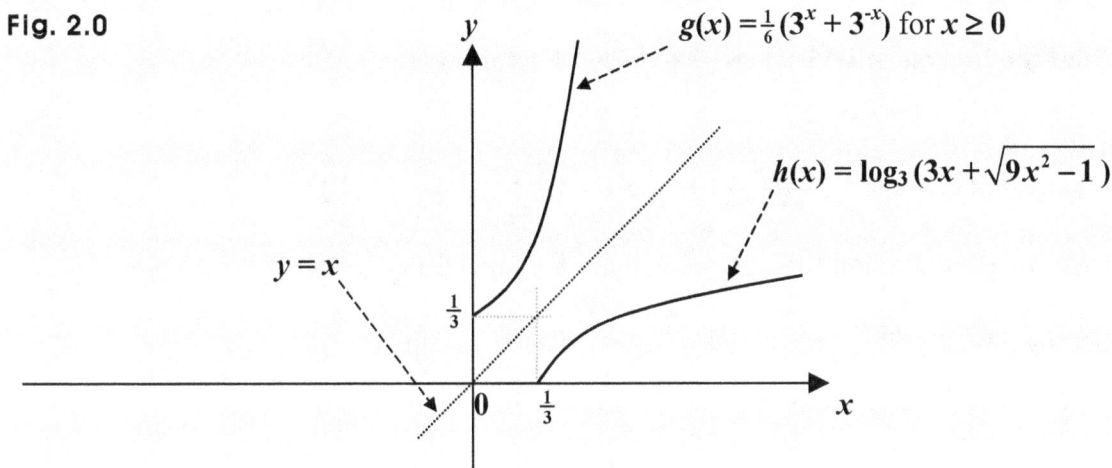

In short:

First, $y = h(x) = \log_3(3x + \sqrt{9x^2 - 1}) \Rightarrow y = \log_3(3x + \sqrt{9x^2 - 1})$.

Next, by the definition for logs: $y = \log_3(3x + \sqrt{9x^2 - 1}) \Leftrightarrow 3x + \sqrt{9x^2 - 1} = 3^y$.

Setting $t = 3^y$, we get: $3x + \sqrt{9x^2 - 1} = t$.

$t - 3x = \sqrt{9x^2 - 1} \Rightarrow (t - 3x)^2 = 9x^2 - 1 \Rightarrow t^2 - 6tx + 9x^2 = 9x^2 - 1 \Rightarrow 6tx = t^2 + 1$

$\Rightarrow x = \frac{t}{6} + \frac{1}{6t}$.

So we get: $x = \frac{1}{6}(t + \frac{1}{t}) = \frac{1}{6}(3^y + \frac{1}{3^y}) = \frac{1}{6}(3^y + 3^{-y})$.

$x \geq 1/3 \Rightarrow 9x^2 - 1 \geq 0 \Rightarrow \sqrt{9x^2 - 1} \geq 0 \Rightarrow 3x + \sqrt{9x^2 - 1} \geq 1$, since $x \geq 1/3$. Thus, we get:

$\log_3(3x + \sqrt{9x^2 - 1}) \geq \log_3 1 = 0 \Rightarrow \log_3(3x + \sqrt{9x^2 - 1}) \geq 0$.

And thus, assuming the inverse is g, we get: $y = g(x) = \frac{1}{6}(3^x + 3^{-x})$ for $x \geq 0$.

Suggestions or Solutions
To the **Problem** in the Example **3**

Find the inverse of $k(x) = \log_3\left(\frac{x}{3} + \sqrt{\frac{x^2-1}{9}}\right) + 1$ for $x \geq 1$.

First, $y = k(x) = \log_3\left(\frac{x}{3} + \sqrt{\frac{x^2-1}{9}}\right) + 1 \Rightarrow y - 1 = \log_3\left(\frac{x}{3} + \sqrt{\frac{x^2-1}{9}}\right)$.

Next, by the definition for logs, we get: $y - 1 = \log_3\left(\frac{x}{3} + \sqrt{\frac{x^2-1}{9}}\right) \Leftrightarrow \frac{x}{3} + \sqrt{\frac{x^2-1}{9}} = 3^{y-1}$.

Setting next, $t = 3^{y-1}$, we get: $\frac{x}{3} + \sqrt{\frac{x^2-1}{9}} = t$. Then, we can get:

$t - \frac{x}{3} = \sqrt{\frac{x^2-1}{9}} \Rightarrow \left(t - \frac{x}{3}\right)^2 = \frac{x^2-1}{9} \Rightarrow t^2 - \frac{2tx}{3} + \frac{x^2}{9} = \frac{x^2-1}{9} \Rightarrow \frac{2tx}{3} = t^2 + \frac{1}{9} \Rightarrow x = \frac{3t}{2} + \frac{1}{6t}$.

So we get: $x = \frac{1}{2}\left(3t + \frac{1}{3t}\right) = \frac{1}{2}\left(3 \cdot 3^{y-1} + \frac{1}{3 \cdot 3^{y-1}}\right) = \frac{1}{2}\left(3^y + 3^{-y}\right)$. Next, getting the range of k,

we get first: $x \geq 1 \Rightarrow x^2 - 1 \geq 0 \Rightarrow \sqrt{\frac{x^2-1}{9}} \geq 0 \Rightarrow \frac{x}{3} + \sqrt{\frac{x^2-1}{9}} \geq \frac{1}{3}$, since $x \geq 1$, so next, we

get: $\log_3\left(\frac{x}{3} + \sqrt{\frac{x^2-1}{9}}\right) \geq \log_3 \frac{1}{3} = -1 \Rightarrow \log_3\left(\frac{x}{3} + \sqrt{\frac{x^2-1}{9}}\right) \geq -1 \Rightarrow \log_3\left(\frac{x}{3} + \sqrt{\frac{x^2-1}{9}}\right) + 1 \geq 0$.

And thus, assuming the inverse is s, we get: $y = s(x) = \frac{1}{2}\left(3^x + 3^{-x}\right)$ for $x \geq 0$.

If not quite sure of the idea behind the processes above, follow the steps below:

To begin with, taking the inverse of an exponential function, what do we get?

We get an exponential function. So this time too, we want to find a function exponential. Functions in both the kinds are inverse of each other, so we can find it by the definition for logs. And the definition is: $y = b^x \Leftrightarrow x = \log_b y$, where $b > 0$, but $b \neq 1$.

And we can put it this way, too: $y = \log_b x \Leftrightarrow x = b^y$, where $b > 0$, but $b \neq 1$.

Getting thus, the inverse either log or exponential, we can apply the definition above to the function we take the inverse of. Then, swapping the variables, we get the inverse.

The definition above is however, just a prototype if you will. So it's in the simplest form. And expressing the definition in general, we can put it the way below:

$$p(y) = \log_b q(x) \Leftrightarrow q(x) = b^{p(y)}, \text{ where } b > 0, \text{ but } b \neq 1.$$

And of course, $p(y)$ is an expression in terms of y, and $q(x)$ is an expression in terms of x.

And for instance, using the definition for logs, we can get:

$$3y + 1 = \log_b (2x + 5) \Leftrightarrow 2x + 5 = b^{(3y + 1)}.$$

So we can get: $3y + 1 = \log_b (2x + 5) \Rightarrow 2x + 5 = b^{(3y + 1)}$, which is an exponential function.

So in this problem, since we need to find the inverse of the log function k, we want to find an exponential function using the function k, along with the definition for logs.

In short, finding the solution to this problem, we want to find a function exponential.

• Now in this problem, we have: $y = k(x) = \log_3 \left(\frac{x}{3} + \sqrt{\frac{x^2-1}{9}}\right) + 1$ for $x \geq 1$.

Then first, we want to put the log function k in the form of $p(y) = \log_b q(x)$.
We can simply do so putting k the way below:

$y = \log_3 \left(\frac{x}{3} + \sqrt{\frac{x^2-1}{9}}\right) + 1 \Rightarrow y - 1 = \log_3 \left(\frac{x}{3} + \sqrt{\frac{x^2-1}{9}}\right)$, which is now in the form above,

since we can take $y - 1$ as $p(y)$, and can take $\frac{x}{3} + \sqrt{\frac{x^2-1}{9}}$ as $q(x)$.

Thus, next, by the definition, we can set: $y - 1 = \log_3 \left(\frac{x}{3} + \sqrt{\frac{x^2-1}{9}}\right) \Leftrightarrow \frac{x}{3} + \sqrt{\frac{x^2-1}{9}} = 3^{y-1}$.

So we now have: $\frac{x}{3} + \sqrt{\frac{x^2-1}{9}} = 3^{y-1}$.   What then, is the next?

Assuming r is the inverse, we need to put the equation above in this form: $x = r(y)$, where $r(y)$ is an expression in terms of y. So we want to isolate x now.   How?

Assuming: $t = 3^{y-1}$, we can get: $\frac{x}{3} + \sqrt{\frac{x^2-1}{9}} = t$.   Then, we can get:

$$t - \frac{x}{3} = \sqrt{\frac{x^2-1}{9}} \Rightarrow (t - \frac{x}{3})^2 = \frac{x^2-1}{9} \Rightarrow t^2 - \frac{2tx}{3} + \frac{x^2}{9} = \frac{x^2-1}{9} \Rightarrow \frac{2tx}{3} = t^2 + \frac{1}{9} \Rightarrow x = \frac{3t}{2} + \frac{1}{6t}.$$

And we know: $t = 3^{y-1}$.   So we get: $x = \frac{1}{2}(3t + \frac{1}{3t}) = \frac{1}{2}(3 \cdot 3^{y-1} + \frac{1}{3 \cdot 3^{y-1}}) = \frac{1}{2}(3^y + 3^{-y})$.

That is, we now have: $x = \frac{1}{2}(3^y + 3^{-y})$.   What then, is the next?

We want to get the range of the original function **k** so that we can get the domain of the inverse. And getting the range, we can begin with the domain of **k**, which is: $x \geq 1$.

Then, we get: $x \geq 1 \Rightarrow x^2 - 1 \geq 0 \Rightarrow \sqrt{\frac{x^2-1}{9}} \geq 0 \Rightarrow \frac{x}{3} + \sqrt{\frac{x^2-1}{9}} \geq \frac{1}{3}$, since $x \geq 1$.

And taking next, the log of both sides, we get:

$$\log_3 (\frac{x}{3} + \sqrt{\frac{x^2-1}{9}}) \geq \log_3 \frac{1}{3} = -1 \Rightarrow \log_3 (\frac{x}{3} + \sqrt{\frac{x^2-1}{9}}) \geq -1 \Rightarrow \log_3 (\frac{x}{3} + \sqrt{\frac{x^2-1}{9}}) + 1 \geq 0.$$

So the range of **k** is: $x \geq 0$, and thus the domain of the inverse is: $y \geq 0$.

And thus, assuming the inverse is **r**, we get: $y = r(x) = \frac{1}{2}(3^x + 3^{-x})$ for $x \geq 0$.

Fig. 3.0

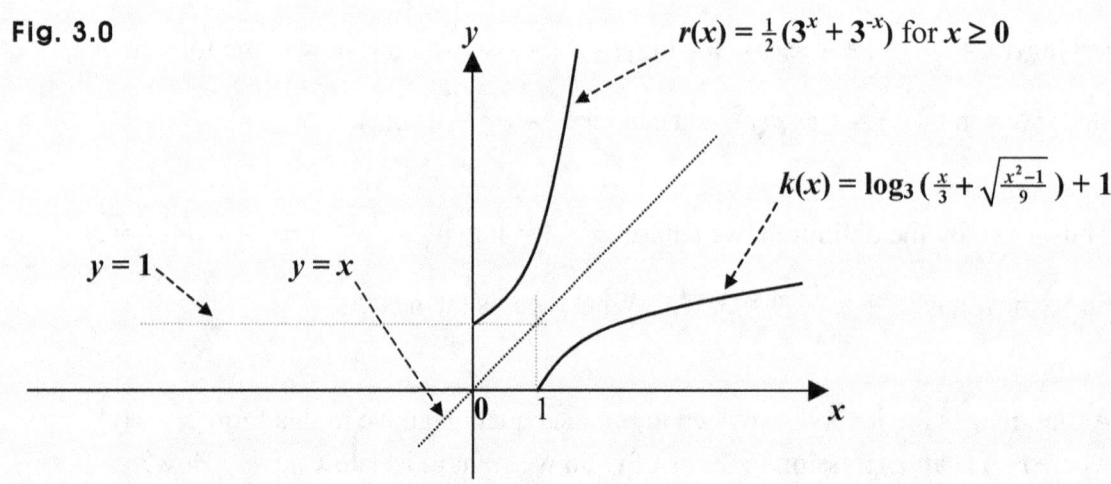

$r(x) = \frac{1}{2}(3^x + 3^{-x})$ for $x \geq 0$

$k(x) = \log_3 (\frac{x}{3} + \sqrt{\frac{x^2-1}{9}}) + 1$

$y = 1$

$y = x$

7. Periodic Functions

Generally, a function produces many different outputs.

In particular, a function one-to-one produces a different output for each input. That is, of such a function, all outputs are different. For any two inputs therefore, the outputs are different. So all outputs are different,.

From a function many-to-one however, same outputs can be produced. That is, from such a function, a particular output can be produced more than once, so some outputs can be the same. And thus, some outputs are the same.

From some functions many-to-one though, many different outputs get produced repeatedly. That is, from such a function, an output gets produced over and over, and the same is true for all the other outputs. And thus, each output repeats.   How?

The same outputs get produced for a group of particular inputs, and the difference between any two inputs in the group is an integer multiple of a particular value.

So for instance, if the particular value is p, and n is an integer, the difference is np. Thus, if $p = 0.3$, the differences can be:

$-2{\cdot}p = -0.6,$ $-1{\cdot}p = -0.3,$ $1{\cdot}p = 0.3,$ $2{\cdot}p = 0.6,$ $3{\cdot}p = 0.9,$ $4{\cdot}p = 1.2,$ etc.

What then, can be the particular inputs?

The particular inputs can be:

-1,   -0.7 = -1 + 0.3,   -0.4 = -0.7 + 0.3,   -0.1 = -0.4 + 0.3,   0.2 = -0.1 + 0.3,

0.5 = 0.2 + 0.3,   0.8 = 0.5 + 0.3,   1.1 = 0.8 + 0.3,   1.4 = 1.1 + 0.3, and so forth.

So for instance, taking the difference between the two inputs **-1** and **-0.1**, we get:

-1 − (-0.1) = -1 + 0.1 = -0.9 = -3·0.3 = -3p, or **-0.1 − (-1) = -0.1 + 1 = 0.9 = 3·0.3 = 3p**.

So the differences among all the particular inputs are integer multiples of a particular value. In other words, the set of all the particular inputs forms an *arithmetic sequence*.

Assuming thus, f is the function described above, and $f(-1) = 2$, we can get:

$f(-1.6) = 2$, $f(-1.3) = 2$, $f(-1) = 2$, $f(-0.7) = 2$, $f(-0.4) = 2$,

$f(-0.1) = 2$, $f(0.2) = 2$, $f(0.5) = 2$, $f(0.8) = 2$,   etc.

That is, we can get:

$f(-1.6) = f(-1.3) = f(-1) = f(-0.7) = f(-0.4) = f(-0.1) = f(0.2) = f(0.5) = f(0.8) = f(1.1)$
$= f(1.4) = f(2) = f(2.6) = f(5.6) = f(6.2) = f(9.5) = f(15.5) = f(16.7) = f(46.7) = \ldots = 2$

And the same is true, too, for all the other outputs from the function f. So for instance:

$f(-1.5) = f(-1.2) = f(-0.9) = f(-0.6) = f(-0.3) = f(0) = f(0.3) = f(0.6) = f(0.9) = f(1.2)$
$= f(1.5) = f(1.8) = \ldots = 3$, for instance.

$f(-1.4) = f(-1.1) = f(-0.8) = f(-0.5) = f(-0.2) = f(0.1) = f(0.4) = f(0.7) = f(1) = f(4.3)$
$= f(4.6) = f(7.9) = \ldots = 1$, for instance.

$f(-1.41) = f(-1.11) = f(-0.81) = f(-0.51) = f(-0.21) = f(0.11) = f(0.71) = f(1.31) = f(4.31)$
$= f(10.31) = f(13.61) = f(22.91) = \ldots = 5$, for instance.

So every time the input gets increased by the particular value p, one same output gets produced *periodically*. What then, should we call such a function?

We can call it a *periodic function*. What then, can we call such a particular value?

We call it a *period*, and thus, the particular value p is the period for the function f. So for instance, the particular value **0.3** is the period for the function f, which is therefore, a periodic function.

Such a period can be taken as an interval in a number line as the x-axis in the x-y plane. So in a function periodic, all the outputs get produced in such an interval. What then, can we say about the range?

The range repeats for every period. Does a periodic function then, have to be one-to-one within a period?

Not necessarily. So it doesn't have to be one-to-one.

The graph of a periodic function can be as follows:

Fig. 7.0

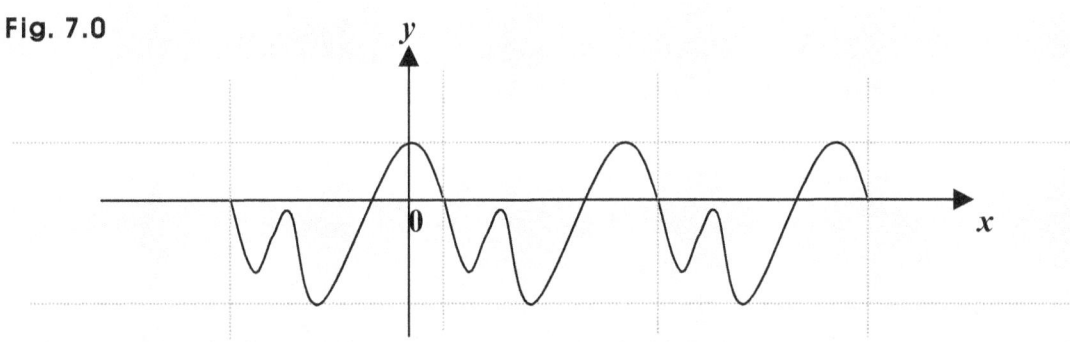

A function periodic can be therefore, many-to-one within a period, too.

That is, some numbers in the range, that is, some outputs can show up more than once within one period. And in particular, if a periodic function is continuous, the same output gets produced at the beginning and at the end of each period, that is, at both ends.

Otherwise, the function cannot be periodic.

If a periodic function is not continuous at the two points each period begins and ends, the two outputs at both ends are different.

So from a periodic function, all the outputs repeat periodically.
More specifically, each and every output equally repeats with the same period.

And thus, in a function periodic, is the period the shortest length between the two inputs for the same output?

No, it's not the case. It's because the function can be many-to-one within the period, too.

Fig. 7.1

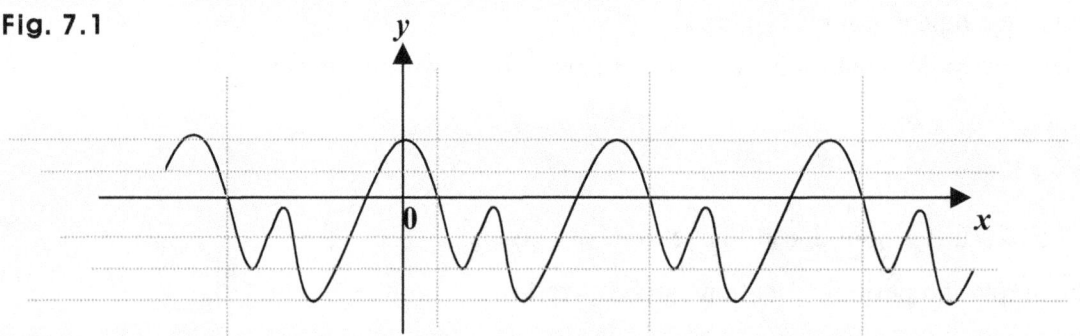

And also, we can see in the graph above, the curve of a periodic function is composed of the same curve segments, each of which is for the period.

What then, is the period?

If we construct the graph of a periodic function, we can see that a particular section of the curve of the function regularly repeats for every constant interval in the x-axis.

Suppose now in the curve, we have found the *smallest section* that repeats continuously for every constant interval in the *x*-axis.

Then, the magnitude of the interval is the period. So we can take the entire curve as a linear collection of duplicates of the smallest section that are put along a line.

We want to note however, that the line stated above is *horizontal*. What do we mean by though, the line horizontal?

The horizontal line is parallel to the axis for the input variable as the *x*-axis in the *x-y* plane. Suppose for instance, the smallest section repeating in a curve is as below:

Fig. 7.2

Next, making duplicates of the section above, we can get:

Fig. 7.3

Now, making a linear collection of the sections putting them along a line horizontal, we can get a periodic curve as below:

Fig. 7.4

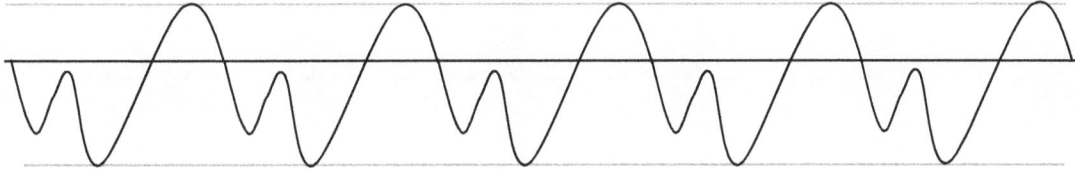

Taking another example, we can get a periodic curve as follows:

Fig. 7.5

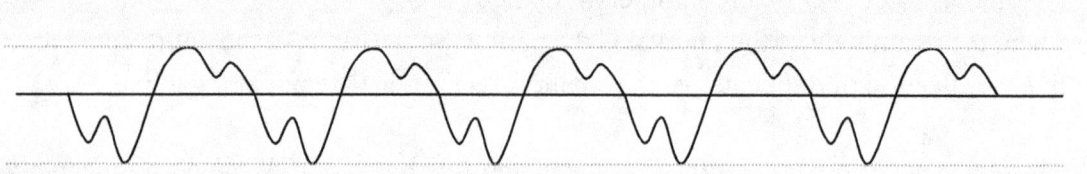

Some curves can look periodic, but they are not from periodic functions. For instance, in the curve below, the same section repeats for every particular interval, but the function is not periodic, because the entire curve is not such a linear collection as described above. The sections are lined up on a line, which is however, not horizontal but slanted. In other words, the sections are connected along a line, not parallel to the *x*-axis.

Fig. 7.6

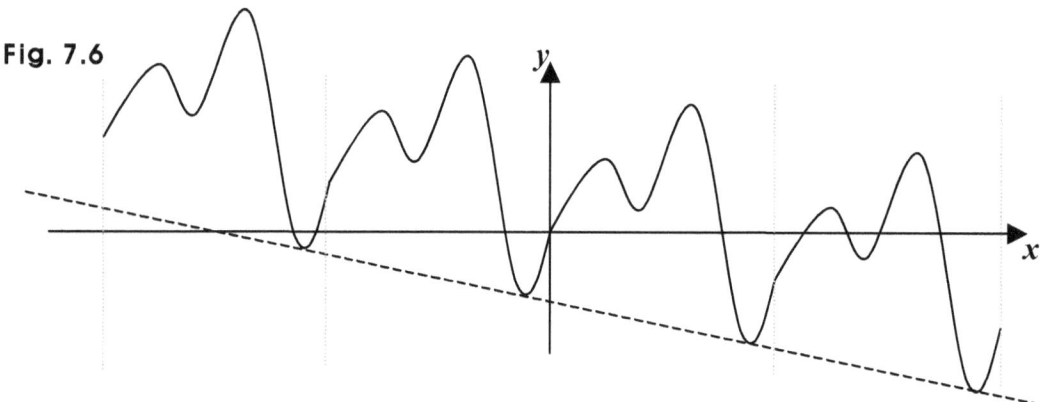

That's because the function is continuous, and the output at the beginning of the interval is not the same as the output at the end of the interval. Assuming for instance, **g** is the function above, and the interval is **p**, we can see in the graph below that **g(p) ≠ g(2p)**.

Fig. 7.7

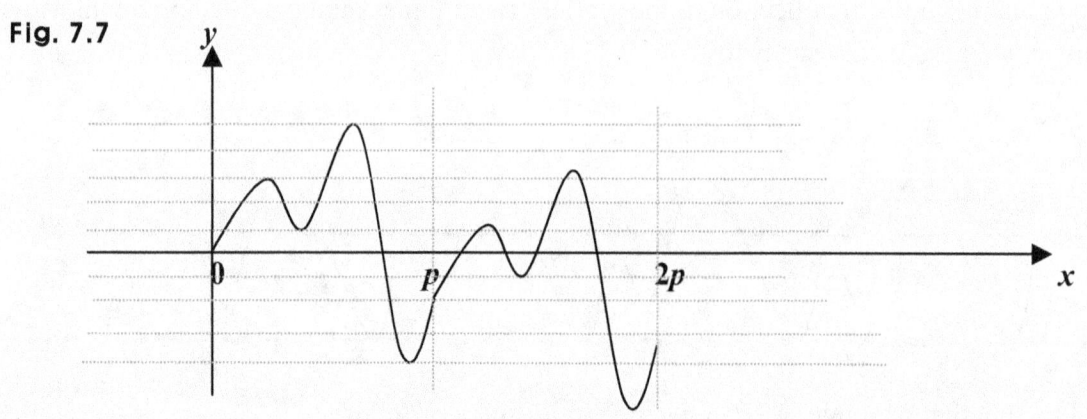

Fig. 7.8

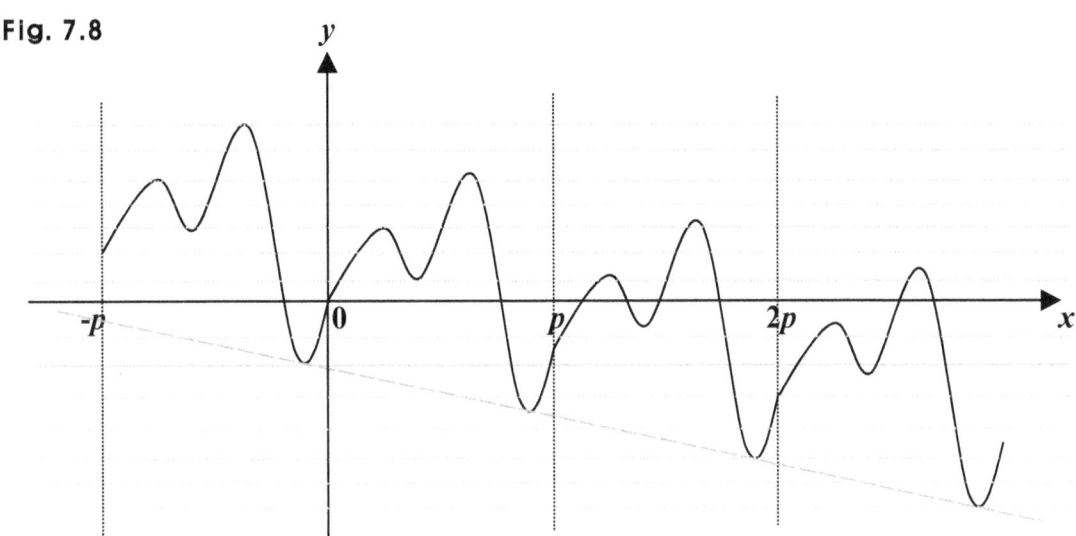

Functions periodic are however, not necessarily continuous. So we can have periodic functions not continuous. Normally though, such functions are continuous within their periods. In other words, such a function is discontinuous at the time only when it moves on to the next period. And examples of such functions periodic and not continuous are:

Fig. 7.9

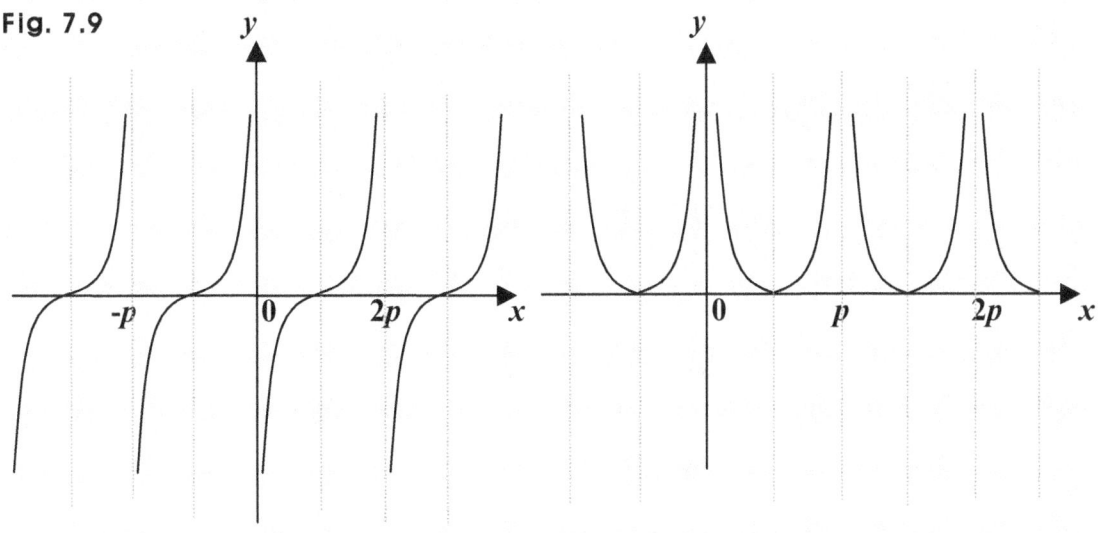

The first of the two above is the graph of a function $y = t(x) = \tan x$ where the period $p = 2\pi$, and the second is the graph of a function $y = T(x) = |\tan x|$ where the period $p = 2\pi$. Such a function as above is called a trigonometric function, a typical function periodic.

How then, can we define a periodic function?

Since the function is periodic, we need to specify the period in the definition. How then, can we specify the period?

From a periodic function, the same output gets produced for particular inputs. And we know the particular inputs form an *arithmetic sequence*.

So let's first, make such a sequence, where the terms are the particular inputs.

Forming an arithmetic sequence, we add a particular value to each term to make the next term.
So for instance, assuming a_0 is the first term, and d is the particular value, we can get an arithmetic sequence called A as follows: $A = \{a_0, a_1, a_2, a_3, ...\}$, where:

$a_1 = a_0 + d, a_2 = a_1 + d, a_3 = a_2 + d, a_4 = a_3 + d$, and so forth.

Thus, the difference between two consecutive terms is the same, and is d in the sequence A. So we call the difference d the *common difference* in the sequence A.

Now, we know all the terms in the sequence A are the particular inputs, for each of which, the same output gets produced.

So what can be the period in the periodic function?

The common difference is the period. How come?

We know all the terms in the sequence A are the particular inputs, for each of which, the same output gets produced. So assuming the periodic function is f, we get:

$f(a_0) = f(a_1) = f(a_2) = f(a_3) = ...$

And we know: $a_1 = a_0 + d, a_2 = a_1 + d, a_3 = a_2 + d, a_4 = a_3 + d, ...$

Thus, we get: $f(a_0) = f(a_0 + d) = f(a_1 + d) = f(a_2 + d) = \ldots$

And assuming 2 is the output for the input a_0, we get:

$2 = f(a_0) = f(a_0 + d) = f(a_1 + d) = f(a_2 + d) = \ldots$

So if we add the period to the current input and take the sum as the next input, the output will be the same. And the same is true, too, for each of all the other outputs.
In other words, every output gets produced periodically.

So if for instance, $y = f(x) = f(x + c)$, what is c?

We have: $f(a_0) = f(a_0 + d) = f(a_1 + d) = f(a_2 + d) = \ldots$, and we can put it this way, too:

$f(a_0) = f(a_0 + d) = f(a_1) = f(a_1 + d) = f(a_2) = f(a_2 + d) = \ldots$

So when $x = a_0$, we get: $f(a_0) = f(a_0 + d)$.   When $x = a_1$, we get: $f(a_1) = f(a_1 + d)$.   And so forth.
And thus, for any value of x, we get: $f(x) = f(x + c)$.   What then, is c?

It is the common difference d, so we get: $c = d$, which is therefore, the period.
So we can define a periodic function f in such a way as follows:

$y = f(x) = f(x + p)$ where p is constant.

From the definition above, we can see that the function f is periodic, and p is the period.
Also, we can see that:

$f(x) = f(x + p) \Rightarrow f(x + p) = f\{(x + p) + p\} = f(x + 2p) \Rightarrow f(x) = f(x + 2p)$

$f(x) = f(x + p) \Rightarrow f(x + 2p) = f\{(x + 2p) + p\} = f(x + 3p) \Rightarrow f(x) = f(x + 3p)$

$f(x) = f(x + p) \Rightarrow f(x + 3p) = f\{(x + 3p) + p\} = f(x + 4p) \Rightarrow f(x) = f(x + 4p)$

. . .

So assuming a function g is periodic, and the period is p, we can set: $g(x) = g(x + np)$ where n is an integer. And of course, the definition of the function g can be as follows:

$y = g(x) = g(x + p)$ where p is constant.

So for instance, if we get: $h(x) = h(x + 3p)$, or $h(x) = h(x - 2p)$, we can see that h is a periodic function where the period is p, since $3p$ and $-2p$ are integer multiples of p.

And for another instance, assuming the period p is 1 in the periodic function $y = f(x)$, we get: $y = f(x) = f(x + 1)$, so we can get:

When $x = 1$, we get: $f(1) = f(1 + 1) = f(2) = f(1 + 1p)$

When $x = 2$, we get: $f(2) = f(2 + 1) = f(3) = f(1 + 2p)$

When $x = 3$, we get: $f(2) = f(3 + 1) = f(4) = f(1 + 3p)$
. . .

When $x = 0.2$, we get: $f(0.2) = f(0.2 + 1) = f(1.2) = f(0.2 + 1p)$

When $x = 2.2$, we get: $f(2.2) = f(2.2 + 1) = f(3.2) = f(0.2 + 3p)$

When $x = 5.2$, we get: $f(5.2) = f(5.2 + 1) = f(6.2) = f(0.2 + 6p)$

. . . $f(5.2) = f(4.2 + 1) = f(4.2) = f(0.2 + 4p) = f(3.2 + 1) = f(3.2) = f(0.2 + 3p)$ . . .

When $x = 17.9$, we get: $f(17.9) = f(17.9 + 1) = f(18.9) = f(0.9 + 18p) = f(0.9)$

When $x = 23.9$, we get: $f(23.9) = f(23.9 + 1) = f(24.9) = f(0.9 + 24p) = f(0.9)$

When $x = -3$, we get: $f(-3) = f(-3 + 1) = f(-2) = f(-3 + 1p)$

When $x = -2$, we get: $f(-2) = f(-2 + 1) = f(-2) = f(-3 + 2p)$

When $x = -1$, we get: $f(-1) = f(-1 + 1) = f(0) = f(-3 + 2p)$
. . .

Thus, in sum, we get: $f(x) = f(x + \text{every multiple of } p)$, where p is the period, of course.

So for instance, we get: $f(0.7) = f(0.7 + np)$ where n is an integer, and p is the period.

And thus, if $p = 1$, we can get: $f(0.7) = f(-0.3)$, because $f(0.7) = f(0.7 + (-1)p) = f(0.7 - 1)$.

However, the same output can be produced for other inputs, too.

That is, a periodic function can be many-to-one for each period.

It is the case if the function periodic is continuous. In other words, if a periodic function is continuous, two or more same outputs can be produced in one period.

So for instance, assuming f is periodic and continuous, and p is the period, we can get not only $f(x) = f(x + p)$ but also $f(x) = f(x + q)$, where q is constant, but is not a period.

For instance, as shown below, $f(x)$ can be 0, too, for x not an integer multiple of p, also.

Fig. 7.A

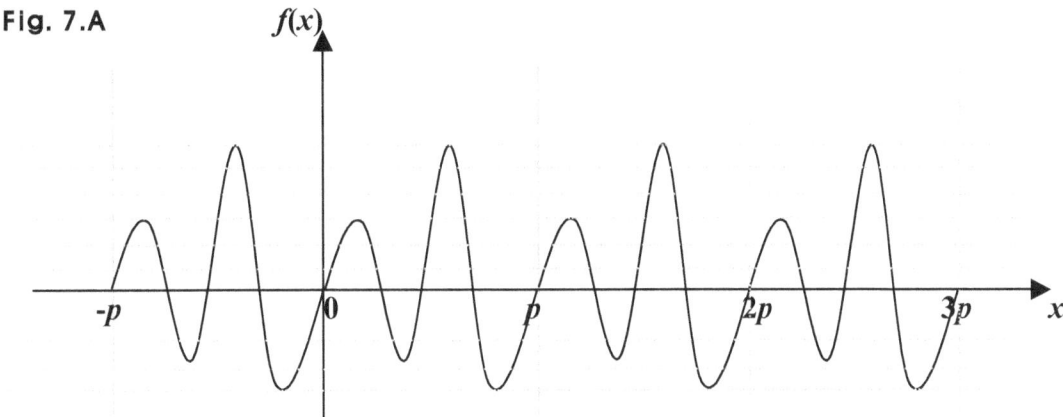

The period is p, and it is *not* the case where: $f(x) = f(x + nq)$ where n is an integer.

That is, we get: $f(x) \neq f(x + nq)$ where q is *not* the period, and n is an integer.

Frequent examples of periodic functions are trigonometric functions, often called briefly trig-functions. Among those, the typical ones are sine, cosine, and tangent functions.

Trig-functions are from trigonometry, which is right triangle geometry, which covers very important ratios, called trigonometric-ratios, and the details are covered in
ALGEBRA EXAMPLES TRIGONOMETRY.

And in fact, many examples of periodic functions are some combinations of the trig-functions, and are wave functions in general. Among such, for instance, we can see functions called Saw tooth function, Square function, and Triangular function.

Examples in Periodic Functions

Doing the examples below, assume that u, v, t, a, and b are nonzero constants.

0.   Assuming $f(x) = f(x + t)$, show that $f(x) = f(x - t)$.

1.   Assuming $f(x) = f(x + u) = f(x + v)$, show that $f(x) = f(x + u + v)$.

2.   Assuming $f(x) = f(x + t)$, show that $f(ax + b)$ is periodic.

3.   Assuming $f(x) = |x - k|$ for $k - 1 \leq x \leq k + 1$, where k is an even integer, find $f(13.2)$ and $f(-6.7)$.

Suggestions or Solutions
To the **Problem** in the Example **0**

Assuming $f(x) = f(x + t)$ where t is a nonzero constant, show that $f(x) = f(x - t)$.

Assuming $s = x + t$ where t is constant, we get: $x = s - t$.

So we get: $f(x) = f(x + t) \Rightarrow f(s - t) = f(s - t + t) = f(s) \Rightarrow f(s) = f(s - t)$.

And thus, we can get: $f(x) = f(x - t)$.

If not quite sure of the idea behind the processes above, follow the steps below:

To begin with, setting $s = x + t$, we get: $x = s - t$, where s is a variable.

How can we though, just set: $s = x + t$?

It's because since t is constant, $(x + t)$ varies as x varies, so $(x + t)$ works like a single variable, and thus, we can replace $(x + t)$ with a letter, and take the letter as a variable. So we can just set: $s = x + t$. And we can get: $x = s - t$.

Thus next, in the equality: $f(x) = f(x + t)$, replacing x with $(s - t)$, we get:

$f(x) = f(x + t) \Rightarrow f(s - t) = f(s - t + t) = f(s) \Rightarrow f(s) = f(s - t)$.

And we know s is a variable, and thus, we can use any other letter as the variable s if the other letter is not used as any other variable in the same expression.   How come?

A variable is like a box that can contain a number at a time, but keeps changing its content. And we use a letter to name the variable if of course, the letter is not used as any other variable in the same expression.

And we know s is a variable, and x is not used as any other variable in the expression where $f(s) = f(s - t)$.   So replacing s with x, we can get: $f(x) = f(x - t)$.

Suggestions or Solutions

To the **Problem** in the Example **1**

Assuming $f(x) = f(x + u) = f(x + v)$ where u and v are nonzero constants, show that $f(x) = f(x + u + v)$.

In the example 0 above, we have: $f(x) = f(x + u) \Rightarrow f(x) = f(x - u)$.

And assuming $s = x - u$, where u is constant, we get: $x = s + u$.

So we get: $f(x) = f(x + v) \Rightarrow f(s + u) = f(s + u + v)$.

And thus, we can get: $f(x + u) = f(x + u + v)$. And we have: $f(x) = f(x + u)$, too.

So we get: $f(x) = f(x + u + v)$.

If not quite sure of the idea behind the processes above, follow the steps below:

First, what do we mean by $f(x) = f(x + u) = f(x + v)$ where u and v are constant and $\neq 0$?

It means that the function f is a periodic function.   What then, is the period?

If we have: $0 < u < v$, then u is the period.   What then, about v?

It is an integer multiple of u.
That is to say that we have: $f(x) = f(x + u) = f(x + nu)$ where n is an integer.
So f is a periodic function where the period is u.

And by the same token, if we have: $0 < v < u$, then v is the period.
So the smaller of the two u and v is the period of the function f.

And next, in the example 0 above, we have a fact below:

- Assuming $f(x) = f(x + t)$ where t is constant and $\neq 0$, we get: $f(x) = f(x - t)$.

So we can get: $f(x) = f(x + u) \Rightarrow f(x) = f(x - u)$.

And assuming $s = x - u$, where u is constant, we get: $x = s + u$.

And we have: $f(x) = f(x + v)$, too.

So since $x = s + u$, if in equality above, replacing x with $(s + u)$, we can get:

$f(x) = f(x + v) \Rightarrow f(s + u) = f(s + u + v)$.

And we know s is a variable, and x is not used as any other variable in the expression where $f(s + u) = f(s + u + v)$. And thus, replacing s with x, we can get:

$f(x + u) = f(x + u + v)$.

And we have: $f(x) = f(x + u)$, too.   So we get: $f(x) = f(x + u + v)$.

In short:

In the example 0, we have: $f(x) = f(x + u) \Rightarrow f(x) = f(x - u)$.

And assuming $s = x - u$, where u is constant, we get: $x = s + u$.

So we get: $f(x) = f(x + v) \Rightarrow f(s + u) = f(s + u + v)$.

And thus, we can get: $f(x + u) = f(x + u + v)$.   And we have: $f(x) = f(x + u)$, too.

So we get: $f(x) = f(x + u + v)$.

Suggestions or Solutions
To the **Problem** in the Example **2**

Assuming *a*, *b*, and *t* are nonzero constants, and $f(x) = f(x + t)$, show that $f(ax + b)$ is periodic.

Suppose $g(x) = f(ax + b)$, and $w = ax + b$.

Then, we can get: $g(x + \frac{t}{a}) = f\{a(x + \frac{t}{a}) + b\} = f(ax + b + t) = f(w + t)$.

So we get: $g(x + \frac{t}{a}) = f(w + t)$. And we have: $f(x) = f(x + t)$. So we get: $f(w + t) = f(w)$.

Thus, we get: $g(x + \frac{t}{a}) = f(w)$.

And we have: $w = ax + b$.   So we get: $f(w) = f(ax + b)$.

And also, we have: $g(x) = f(ax + b)$.   So we get: $g(x) = f(w)$.

Besides, we have this, too: $g(x + \frac{t}{a}) = f(w)$. So we get: $g(x + \frac{t}{a}) = g(x)$.

And thus, $g(x)$ is a periodic function where the period is: $\frac{t}{a}$.

Now, we have: $g(x) = f(ax + b)$. So $f(ax + b)$ is periodic, too, since g is periodic.

If not quite sure of the idea behind the processes above, follow the steps below:

To begin with, we know f is periodic, since we have: $f(x) = f(x + t)$ where t is constant.

Suppose now, $g(x) = f(ax + b)$, and $w = ax + b$.   Why setting: $g(x) = f(ax + b)$, though?

That's the trick. In this example, we want to show that $f(ax + b)$ is periodic.

So showing $g(x)$ is periodic, we show $f(ax + b)$ is periodic, too.

Then, we can get: $g(x + \frac{t}{a}) = f\{a(x + \frac{t}{a}) + b\} = f(ax + b + t) = f(w + t)$.

Why setting: $g(x + \frac{t}{a}) = f\{a(x + \frac{t}{a}) + b\}$, though?

That's another trick. We will get to use: $f(x) = f(x + t)$, which means f is periodic.

So now, getting back to: $g(x + \frac{t}{a}) = f\{a(x + \frac{t}{a}) + b\} = f(ax + b + t) = f(w + t)$, we get:

$g(x + \frac{t}{a}) = f(w + t)$. And we have: $f(x) = f(x + t)$.   So we get: $f(w + t) = f(w)$.

Thus, we get: $g(x + \frac{t}{a}) = f(w)$.

And we have: $w = ax + b$.   So we get: $f(w) = f(ax + b)$.

And also, we have: $g(x) = f(ax + b)$.   So we get: $g(x) = f(w)$.

Besides, we have this, too: $g(x + \frac{t}{a}) = f(w)$.   So we get: $g(x + \frac{t}{a}) = g(x)$.
And we know $\frac{t}{a}$ is constant, since a and t are constant.

So $g(x)$ is periodic, because from g, the same output repeats at every $\frac{t}{a}$.
That is, $g(x)$ is a periodic function where the period is: $\frac{t}{a}$.

Now, we have: $g(x) = f(ax + b)$.   So $f(ax + b)$ is periodic, too, since g is periodic.

In fact, if $f(ax) = f\{a(x + \frac{b}{a})\}$, then $f(ax)$ is periodic, and the period can be said to be $\frac{b}{a}$.

So the period of $f(ax + b)$ can be said to be $\frac{b}{a}$, too.

That's because we can get: $f(ax + b) = f\{a(x + \frac{b}{a})\}$.

Besides, if $f(x)$ is periodic, then $f(ax)$ is periodic, too. And vice versa.

Suggestions or Solutions
To the **Problem** in the Example **3**

Assuming $f(x)$ is periodic, the period is 2, and $f(x) = |x - k|$ for $k - 1 \leq x \leq k + 1$, where k is an even integer, find $f(13.2)$ and $f(-6.7)$.

What do we mean by this: $f(x) = |x - k|$ for $k - 1 \leq x \leq k + 1$, where k is even?

When $k = -4$, $f(x) = |x + 4|$ for $-5 \leq x \leq -3$.

When $k = -2$, $f(x) = |x + 2|$ for $-3 \leq x \leq -1$.

When $k = 0$, $f(x) = |x|$ for $-1 \leq x \leq 1$.

When $k = 2$, $f(x) = |x - 2|$ for $1 \leq x \leq 3$.

When $k = 4$, $f(x) = |x - 4|$ for $3 \leq x \leq 5$.

And so forth.

So putting $f(x)$ in a graph, we can put it the way below:

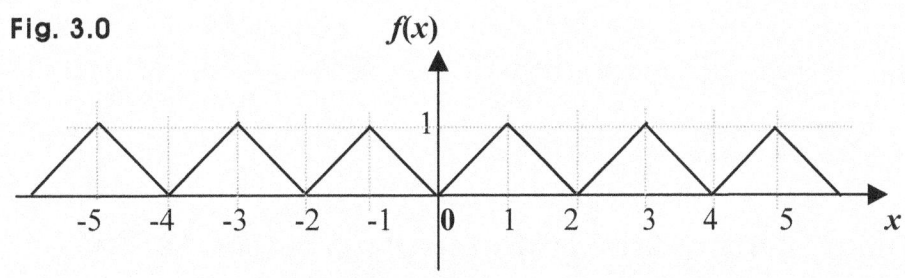

Fig. 3.0

In the graph above, we can see that the period is 2.

Now, if we need to get $f(13.2)$, the interval has to be: $13 \leq x \leq 15$.

And we have: $k - 1 \leq x \leq k + 1$, where k is even. So we get: $k - 1 = 13 \Rightarrow k = 14$.

Then, we get: $f(x) = |x - 14|$ for $13 \leq x \leq 15$.

So we get: $f(13.2) = |13.2 - 14| = 0.8$.

And next, if we need to get $f(-6.7)$, the interval has to be: $-7 \leq x \leq -5$.

So we get: $k - 1 = -7 \Rightarrow k = -6$.

Then, we get: $f(x) = |x + 6|$ for $-7 \leq x \leq -5$.

So we get: $f(-6.7) = |-6.7 + 6| = 0.7$.

And we can get the solution the way below, too:

Since f is periodic and the period is 2, we can set: $f(x) = f(x + 2)$.

And also, we can set: $f(x) = f(x - 2)$, since f is periodic and the period is 2.

And since the period is 2, we can get: $f(x) = f(x + 2n)$ for n integer.

And in this case, we can use the interval as follows: $-1 \leq x \leq 1$.

Then, we need to use: $f(x) = |x|$ for $-1 \leq x \leq 1$.

Then, we can get:

$f(x) = f(x - 2 \cdot 7) = f(x - 14) \Rightarrow f(13.2) = f(13.2 - 14) = f(-0.8) = |-0.8| = 0.8$.

$f(x) = f(x + 2 \cdot 3) = f(x + 6) \Rightarrow f(-6.7) = f(-6.7 + 6) = f(-0.7) = |-0.7| = 0.7$.

www.ingramcontent.com/pod-product-compliance
Lightning Source LLC
Chambersburg PA
CBHW082034190526
45165CB00020B/2501